Lecture Notes in Statistics   108

Edited by P. Bickel, P. Diggle, S. Fienberg, K. Krickeberg, I. Olkin, N. Wermuth, S. Zeger

Springer
New York
Berlin
Heidelberg
Barcelona
Budapest
Hong Kong
London
Milan
Paris
Santa Clara
Singapore
Tokyo

Wesley L. Schaible
(Editor)

Indirect Estimators in
U.S. Federal Programs

Springer

Wesley L. Schaible
Bureau of Labor Statistics
2 Massachusetts Avenue Northeast
Room 4915
Washington, D.C. 20212

Library of Congress Cataloging-in-Publication Data Available
Printed on acid-free paper.

Camera ready copy provided by the author.
Printed and bound by Braun-Brumfield, Ann Arbor, MI.
Printed in the United States of America.

9 8 7 6 5 4 3 2 1

ISBN 0-387-94616-0 Springer-Verlag New York Berlin Heidelberg

PREFACE

This monograph is a result of the collaborative efforts of members of the Subcommittee on Small Area Estimation sponsored by the U.S. Federal Committee on Statistical Methodology. The Federal Committee is concerned with issues affecting Federal statistics and conducts its work through subcommittees organized to study particular issues. The Subcommittee on Small Area Estimation consisted of Robert Fay, Joe Fred Gonzalez, Linnea Hazen, William Iwig, John Long, Donald Malec, Wesley Schaible (chair), and Alan Tupek. Subcommittee members were responsible for authoring chapters and, except for the introductory and concluding chapters, authors are indicated with the chapter title. The introductory and concluding chapters were authored by the Subcommittee Chair; however, they resulted from discussions which included the entire Subcommittee as well as other interested parties.

The purpose of this monograph is to document, in a manner that facilitates comparisons, the practices and estimation methods of federal statistical programs that use indirect estimators to produce published estimates. These programs are generally formed to respond to legislative requirements or, alternatively, to subnational data needs. Eight programs in the U.S. Federal Statistical System were identified and included in this report: state infant and maternal health characteristics (National Center for Health Statistics); state and county estimates of personal income, annual income, and gross product (Bureau of Economic Analysis); postcensal population estimates for counties (Bureau of the Census); employment and unemployment for states (Bureau of Labor Statistics); cotton, rice, and soybean acreage for counties (National Agricultural Statistics Service); livestock inventories, crop production, and acreage for counties (National Agricultural Statistics Service); disabilities, hospital utilization, physician and dental visits for states (National Center for Health Statistics); and state estimates of median income for four person families (Bureau of the Census).

Throughout the preparation of this report, a number of reviewers assisted the Subcommittee. The Subcommittee thanks Maria Gonzalez, Chair of the Federal Committee on Statistical Methodology, and Committee members for their contributions throughout the development and preparation of the report. The Subcommittee also thanks Alan Dorfman, Steve Miller, and especially Robert Casady, all of the Bureau of Labor Statistics, for helpful discussions and comments. In addition, the Subcommittee extends its thanks to Gordon Brackstone and the staff of Statistics Canada, to Wayne Fuller of Iowa State University, and especially to Graham Kalton of Westat, Inc. for valuable comments and the time so generously provided to review the report. We would also like to thank Francis Kalton for invaluable help in dealing with software specific notation problems.

Wesley L. Schaible

CONTENTS

CONTENTS

CHAPTER 1

Introduction and Summary

1.1 Introduction

Federal statistical agencies produce estimates of a variety of population quantities for both the nation as a whole and for subnational domains. Domains are commonly defined by demographic and socioeconomic variables. However, geographic location is perhaps the single variable used most frequently to define domains. Regions, states, counties, and metropolitan areas are common geographic domains for which estimates are required. Federal agencies use different data systems and estimation methods to produce domain estimates. Those systems designed for the purpose of producing published estimates use standard, direct estimation methods. Data systems are designed within time, cost and other constraints which restrict the number of estimates that can be produced by standard methods. However, the demand for additional information and the lack of resources to design the required data systems have led federal statistical agencies to consider non-standard methods. Estimation methods of a particular type, referred to as small area or indirect estimators, have sometimes been used in these situations.

The U.S. Federal Committee on Statistical Methodology was organized to investigate methodological issues in the production of federal statistics. The committee conducts its work through subcommittees organized to study particular issues. In April 1991, a subcommittee was charged with the task of identifying and documenting federal statistical programs in the U.S. that have used indirect estimators for the production of published estimates. The report that resulted, Indirect Estimators in Federal Programs (U.S. Office of Management and Budget 1993), documents eight programs and provides discussion on the definition and characteristics of indirect estimators. In general, this monograph is an edited and slightly expanded version of the Federal Committee report; this chapter also incorporates additional material from Schaible (1993).

The purpose of the Federal Committee Report and this monograph is to document, in a manner that will facilitate comparisons, the practices and estimation methods of the U.S. federal statistical programs that use indirect estimators. Only programs that use indirect estimators for the production of published estimates are included; whether a data system is based on a census (including administrative records) or a sample survey has no bearing on the inclusion of a program. The focus of this report is on the method by which estimates are produced. The methods and practices of eight programs are documented here; three are located in the Department of Commerce, two in the Department of Agriculture, two in the Department of Health and Human Services, and one in the Department of Labor. Other applications of indirect estimators occur in federal statistical agencies but descriptions of these applications have not been included in this report. Most of these methods were not included because they were not used to produce published estimates. This publication restriction, a somewhat arbitrary indicator of program importance, keeps the number of programs included to a manageable level but leads to the omission of other interesting methods (for example, Fay and Herriott 1979).

This introductory chapter includes brief discussions of small area estimation terminology; definitions of direct and indirect estimators; some characteristics of indirect estimators; and summary descriptions of the programs included in the report. Each program is documented, following a standard format, in the individual chapters of the report. The intent is to create program descriptions that will not only provide complete, self-contained documentation for each individual program but also facilitate comparisons among programs. Although the focus of the report is on estimation methods, the description of each program includes material on program history, policies, evaluation practices, estimation methods, and current problems and activities. In addition to the standard chapter format, attempts have been made to employ common notation throughout the chapters to facilitate comparisons of estimation methods. The report concludes with a number of recommendations and cautions.

1.1.1 Geographic Subdivisions of the United States

There is considerable demand for statistical data relating to geographic subdivisions of the United States. It should be noted that there is substantial variation in both geographic area and population size among these domains. For example, according to the 1990 census of the population, one eighth of the total U.S. population resides in one state, California. The largest state, Alaska, is over 400 times the size, but yet has only half the population, of the smallest state, Rhode Island. States are further divided into over 3000 counties. More people reside in the county with the largest population, Los Angeles County, than in 43 of the 50 states. The variability in population size among the states, counties and other geographic areas adds to the difficulty of making geographical estimates from national, survey based data systems.

1.1.2 Terminology

Terms used to describe indirect estimators can be confusing. Increased interest in non-traditional estimators for domain statistics has occurred recently among survey statisticians and, even though the term "small area estimator" is commonly used, uniform terminology has not yet evolved. This term is frequently used because in most applications of these estimators the domains of interest have been geographic areas. However, the word "small" is misleading. It is the small number of sample observations and the resulting large variance of standard direct estimators that is of concern, rather than the size of the population in the area or the size of the area itself. The word "area" is also misleading since these methods may be applied to any arbitrary domain, not just those defined by geographic boundaries. Other terms used to describe these estimators include "area breakdowns" (Woodruff 1966), "local area" (Ericksen 1974), "small domain" (Purcell and Kish 1979), "subdomain" (Laake 1979), "small subgroups" (Holt, Smith, and Tomberlin 1979), "subprovincial" (Brackstone 1987), "indirect" (Dalenius 1987), and "model-dependent" (Särndal 1984). The term "synthetic estimator" has also been used to describe this class of estimators (NIDA 1979) and, in addition, to describe a specific indirect estimator (NCHS 1968). Survey practitioners sometimes refer to indirect estimators as "model-based" whereas this term is rarely, if ever, used to describe direct estimators. However, direct estimators can be motivated by and justified under models as readily as indirect estimators.

There is also lack of agreement on what to call the class of direct estimators. In addition to "direct" (Royall 1973), authors have used "unbiased" (Gonzalez 1973), "standard" (Holt, Smith, and Tomberlin 1979), "valid" (Gonzalez 1979), and "sample-based" (Kalton 1987). In the remainder of this paper, the words "direct" and "indirect" will be used to describe traditional and small area estimators, respectively.

1.2 Direct and Indirect Estimators

Perhaps the most common measure of error of an estimator is the mean square error, composed of the sum of the variance of the estimator and the squared bias of the estimator. Biases can rarely be estimated with any degree of confidence. If an estimator is unbiased or approximately unbiased, the variance of the estimator, which can be estimated from the available data, is a satisfactory measure of error of the estimator. This leads to the selection of estimators that are unbiased or approximately unbiased in most applications. Such estimators allow data systems to be designed so that estimates with a predictable level of error can be produced with high probability and, in addition, estimated measures of error can be provided to accompany estimates.

Federal statistical programs are generally designed using direct estimators which are unbiased, or approximately unbiased, under finite population sampling theory. Samples are often used and, given adequate resources, the sample design specifies population and domain sample sizes large enough to produce direct estimates that meet reliability requirements for the survey. When a domain sample size is too small to make a reliable domain estimate using the direct estimator, a decision must be made whether to produce estimates using an alternative procedure. The alternative estimators considered are those that increase the effective sample size and decrease the variance by using data from other domains and/or time periods through models that assume similarities across domains and/or time periods. These estimators are generally biased, but if the mean square error of the alternative estimator can be demonstrated to be small compared to the variance of the direct estimator, the selection of the alternative estimator may be justified. In extreme situations, there may be no sample units in the domain of interest and, if an estimate is to be produced, an alternative estimator will be required.

1.2.1 Definitions

Indirect estimators have been characterized in the Bayesian and empirical Bayes literature as estimators that "borrow strength" by the use of values of the variable of interest from domains other than the domain of interest. This approach can be used to provide a working definition of direct and indirect estimators for a broad class of population quantities including means and totals. A *direct estimator* uses values of the variable of interest only from the time period of interest and only from units in the domain of interest. An *indirect estimator* uses values of the variable of interest from a domain and/or time period other than the domain and time period of interest. Three types of indirect estimators can be identified. A *domain indirect estimator* uses values of the variable of interest from another domain but not from another time period. A *time indirect estimator* uses values of the variable of interest from another time period but not from another domain. An estimator that is both *domain and time indirect* uses values of the variable of interest from another domain and another time period.

Indirect estimators depend on values of the variable of interest from domains and/or time periods other than that of interest. These values are brought into the estimation process through a model that, except in the most trivial case, depends on one or more auxiliary variables that are known for the domain and time period of interest. To the extent that applicable models can be identified and the required auxiliary variables are available, indirect estimators can be created to produce estimates. Perhaps the simplest example of an indirect estimator is the use of the sample mean of the entire sample as the estimator for a specific domain. For example, the use of the mean from a national sample as an estimate for a particular state. To the extent that information related to the variable of interest is available for the state, an indirect estimator which is "better" than the national mean can be defined. The availability of auxiliary variables and an appropriate model relating the auxiliary variables to the variable of interest are crucial to the formation of indirect estimators. However, the definition of direct and indirect estimators does not depend on whether or not auxiliary variables from outside the domain or time period of interest are used.

The clear distinction between direct and indirect estimators made in the discussion above reflects the situation during the design stage of a data system. However, when estimators reflect the realities associated with data system implementation, the distinction becomes a little less clear. For example, nonresponse is a common problem in data collection efforts. To the extent that nonresponse occurs, even direct estimators must rely on model-based assumptions relating the known information for responders to the unknown information for nonresponders. Even though they will not be discussed here, secondary estimation methods such as nonresponse adjustment, raking, and seasonal adjustment borrow strength and are subject to some of the same concerns as basic indirect estimators.

1.2.2 Characteristics of Indirect Estimators

Insight into the differences between direct and indirect estimators may be gained by inspecting their underlying models. Notation will be required. Let

$d = 1, 2, \ldots, D$ denote domains,
$t = 1, 2, \ldots, T$ denote time periods,
$i = 1, 2, \ldots, N_{dt}$ denote units in the population at time t and in domain d,

and

Y_{dti} denote the variable of interest associated with unit/observation dti.

In addition, within the domain and time period of interest, let s_{dt} denote the set of units that are in the sample and $\tilde{s}_{dt}$, the set of units not in the sample.

Table 1. Direct and Indirect Estimators of Model Parameters and the Finite Population Mean, $\overline{Y}_{dt}$, for the Family of Models Defined by $E(Y_{dti}) = \mu_{dt}$

Expectation Model, $E(Y_{dti})$	BLUE for the Model Parameter	BLUP for $\overline{Y}_{dt}$	Type of Estimator
μ_{dt}	$\hat{\overline{Y}}_{dt}$	$(1/N_{dt})\left(\sum_{s_{dt}} Y_{dti} + \sum_{\tilde{s}_{dt}} \hat{\overline{Y}}_{dt} \right)$	Direct
$\mu_{.t}$	$\hat{\overline{Y}}_{.t}$	$(1/N_{dt})\left(\sum_{s_{dt}} Y_{dti} + \sum_{\tilde{s}_{dt}} \hat{\overline{Y}}_{.t} \right)$	Domain Indirect
$\mu_{d.}$	$\hat{\overline{Y}}_{d.}$	$(1/N_{dt})\left(\sum_{s_{dt}} Y_{dti} + \sum_{\tilde{s}_{dt}} \hat{\overline{Y}}_{d.} \right)$	Time Indirect
$\mu_{..}$	$\hat{\overline{Y}}_{..}$	$(1/N_{dt})\left(\sum_{s_{dt}} Y_{dti} + \sum_{\tilde{s}_{dt}} \hat{\overline{Y}}_{..} \right)$	Domain and Time Indirect

The example in Table 1 illustrates several points that help to better understand characteristics of and the relationship between direct and indirect estimators.

1. A domain and time specific model defines a family of models. For example, associated with the single parameter, domain and time specific model, $E(Y_{dti}) = \mu_{dt}$ are three other models. With appropriate independence and variance assumptions, each model leads to a best linear unbiased estimator (BLUE) for the model parameter and a best linear unbiased predictor (BLUP) for the population mean. The domain and time specific model leads to a direct estimator whereas the three remaining models lead to a domain indirect, a time indirect, and a domain and time indirect estimator.

2. If the Y's are independent with constant variance, then the BLUE's for the parameters of the four models in this family are: 1) the sample mean in the domain and time period of interest for the model parameter, μ_{dt}, 2) the sample mean for

the specified time period across all domains for the model parameter, $\mu_{.t.}$, 3) the sample mean for the specified domain across all time periods for the model parameter, $\mu_{d..}$, and 4) the sample mean across all domains and time periods for the model parameter, $\mu_{...}$.

3. The objective in finite population estimation problems is not to estimate a model parameter, but rather to estimate the population mean (or total) for a particular domain and time period. Within the domain and time of interest, the BLUP of the population total is obtained by adding the known sum of the values for sampled units to the estimated sum of the unobserved values for the nonsampled units. In this example, the unobserved value associated with each nonsampled unit is estimated by the BLUE for the corresponding model parameter. The BLUP for the population mean is obtained by dividing the predicted total by the number of units in the population.

4. For the domain and time specific model, the BLUE for the model parameter is algebraically equivalent to the BLUP for the finite population parameter. For the remaining models in the family, the BLUE for the model parameter and the BLUP for the finite population parameter are not the same. In these cases, the BLUE for the model parameter is an unbiased predictor for the finite population mean, but it is not the BLUP.

5. It is straightforward to verify that the direct estimator is robust against model failure in the sense that it is unbiased, not only under the domain and time specific model, but under each of the models in the family. Indirect estimators are not robust in the same sense; each of the indirect estimators in the family is biased under the domain and time specific model. However, the domain indirect and the time indirect estimators are more robust against model failure than the domain and time indirect estimator in the sense that they are unbiased, not only under the model that leads to each estimator, but also under the $E(Y_{dti}) = \mu_{...}$ model that leads to the domain and time indirect estimator. Without evidence to the contrary, the domain and time specific model will be the most plausible in the family, and the bias of indirect estimators under this model will continue to be a major source of concern surrounding applications of indirect estimators.

6. This simple example can also be used to help understand the importance of keeping the purpose of the analysis in mind when selecting an indirect estimator. Not all indirect estimators will be equally appropriate for a given analysis. For example, if the purpose of the analysis is to make comparisons across domains for a given time period, it would serve no purpose to use the domain indirect estimator above since this estimator would produce essentially the same estimate for every

domain. Even though this is an extreme example, the point is clear. Domain indirect estimators are based on models that assume the expectation of the variable of interest is the same across domains with respect to some model parameter. This inconsistency between the purpose of the analysis and the method used to produce estimates will be avoided if a time indirect estimator is utilized. If, instead of making comparisons across domains, the purpose of the analysis is to make comparisons across time periods within a given domain, it may be appropriate to select from among the domain indirect estimators. However, it should be stressed that, in practice, the performance of both domain and time indirect estimators depends on the available information and how accurately the model that incorporates this information depicts the actual application of interest.

In addition to the characteristics illustrated in the example above, there are several other, fairly well-known characteristics of indirect estimators that are important to keep in mind.

   • Since they not only incorporate observations from the domain and time period of interest, but also from other domains and/or time periods, indirect estimators have smaller variances than the direct estimator in the same family. Holt, Smith, and Tomberlin (1979) discuss estimation of the variance of the modified (best linear unbiased) synthetic estimator and Royall (1979) presents variances of several indirect estimators resulting from various prediction models. Care must be taken since the variance alone may not lead to valid confidence intervals. See, for example, Räbäck and Särndal (1982) and Särndal and Hidiroglou (1989).

   • Generally, a meaningful measure of error is difficult to produce for an indirect estimator for a specific domain and time. An indirect estimator will be biased if the model assumptions leading to the estimator are not satisfied, and the magnitude of the bias is likely to vary with each application. Estimation of biases is, of course, difficult. Gonzalez and Waksberg (1973) consider the problem of estimating the mean squared error of synthetic estimators, and Prasad and Rao (1990) discuss the estimation of the mean squared error of indirect estimators. Care must be taken when interpreting estimated mean squared errors of indirect estimators; some approaches provide an average measure over all domains rather than an individual measure associated with a specific domain.

   • For a given application and estimator, biases in different domains will differ since the model will likely be a better representation of reality in some domains than in others. Many indirect estimators produce domain estimates whose distribution has smaller variance than the corresponding distribution of domain population values being estimated. That is, when domain population values are close to the average population value, indirect estimators have relatively small biases. However, when domain population values are not close to the overall population value, indirect estimators tend to have relatively large biases which act in such a way to make the estimates closer to the average population value. There is considerable evidence illustrating this characteristic (Gonzalez and Hoza 1978; Schaible et al. 1977 and 1979; and Heeringa 1981). Not all indirect estimators display this characteristic to the same extent. Spjøvoll and Thomsen (1987), Lahiri (1990), and Ghosh (1992) have recently addressed this problem and suggest constrained approaches.

1.3 Organization of Program Chapters

As discussed in the previous section, indirect estimators borrow strength and can be classified into three types: domain indirect, time indirect, and domain/time indirect. In addition to this classification, indirect estimators are commonly expressed in different forms, that is, different algebraic expressions. Each of the eight programs described in this report uses one of the following three common indirect estimators: synthetic, regression, or composite. The order of chapters describing programs follows this classification of estimators. That is, the program that uses a synthetic estimator is presented first in Chapter 2, followed by the programs that use regression estimators in Chapters 3 through 6; those programs that use composite estimators are presented in Chapters 7 through 9. Some of the programs have used different estimators at different times; however, emphasis is placed on the estimator that was last used to publish estimates.

As with all indirect estimators, synthetic estimators may be domain indirect, time indirect, or domain and time indirect. For example, a domain indirect synthetic estimator for a population total in domain d and time t may be written as

$$\hat{T}_{(SYN),d,t} = \sum_{h=1}^{H} N_{dth}\hat{\bar{Y}}_{.th} \, ,$$

where $h = 1, 2, \ldots, H$ denotes poststrata; N_{dth} denotes the number of population units in domain d, time t, and poststratum h; and $\hat{\bar{Y}}_{.th}$ denotes the sample mean across all domains for time t and poststratum h. A domain indirect synthetic estimator used by the National Center for Health Statistics is described in Chapter 2.

Regression estimators may be direct or, like the synthetic estimator, domain indirect, time indirect, or domain and time indirect depending on how the parameters are estimated. For example, a domain indirect regression estimator for a population total may be written as

$$\hat{T}_{(REG),d,t} = \sum_{i=1}^{N_{dt}} \mathbf{x}_{dti}\hat{\beta}_{.t} \, ,$$

where N_{dt} denotes the number of units in population dt, $i = 1,2, \ldots , N_{dt}$ denotes units within population dt, $\mathbf{x}_{dti}$ denotes a row vector of known auxiliary variables and $\hat{\beta}_{.t}$, a column vector of estimators of the corresponding regression coefficients. The regression coefficients are estimated using y values from one or more domains besides d but within the time period t. The four programs described in Chapters 3 through 6 use indirect regression estimators to produce estimates.

Although the synthetic estimator is discussed here as a separate type of estimator, it can be written as a special case of a regression estimator where the auxiliary variables are defined to be variables indicating whether or not each unit is in poststratum h or not. More precisely, a domain indirect synthetic estimator with H poststrata can be expressed as a domain indirect regression estimator with H summed terms. Each term corresponds to a poststratum (h) and is the sum of the N_{dt} products of 1) an auxiliary variable which takes the value one when unit i is in poststratum h and zero otherwise and 2) the estimated $\hat{\beta}_{.th}$. Assuming simple random sampling and using ordinary least squares to estimate the regression parameters, each estimated $\hat{\beta}_{.th}$ is $\hat{\overline{Y}}_{.th}$. Since $\hat{\overline{Y}}_{.th}$ is a constant with respect to the summation over i, the sum of the N_{dt} auxiliary indicator variables indicating that unit i is in stratum h is simply N_{dth}, and the sum of the H terms in the regression estimator is seen to be identical to the synthetic estimator expression above.

Another interesting special case of indirect regression estimation occurs when there is only one auxiliary variable and the sum of the population y values is known for the entire population, but the sums for the domain populations are not known. The domain indirect regression (ratio) estimator in this case can be written as,

$$\hat{T}_{(REG),d,t} = \frac{X_{dt.}}{X_{.t.}} Y_{.t.} \, ,$$

where $X_{dt.}$ denotes the sum of the auxiliary variable over units in domain d and time t, $X_{.t.}$ denotes the sum of the auxiliary variable over units and domains in time t, and $Y_{.t.}$ denotes the sum of the variable of interest over units and domains in time t.

This estimator is interesting in that the uncertainty associated with it is not in any way due to sampling. In fact, it can be argued that since there is no sample, this is not an estimator. However, under superpopulation approaches to finite population sampling, it can be considered as an estimator and is unbiased under the model, $EY_{dti} = x_{dti}\beta_{.t}$. A version of this estimator used by the Bureau of Economic Analysis is described in Chapter 3.

A composite estimator may be written as,

$$\hat{T}_{(COM),d,t} = w_{dt}\hat{T}_1 + (1 - w_{dt})\hat{T}_2 \ ,$$

where w_{dt} is a weight, usually between zero and one, and $\hat{T}_1$ and $\hat{T}_2$ are component estimators. Typically, in small area estimation applications, one component estimator is direct and the other is either domain or time indirect. Note that requiring a component estimator to be direct necessitates that at least one observation be available from the domain of interest. Synthetic and indirect regression estimators can be used even if there are no observations from the domain of interest. There are a variety of approaches to defining the weight w_{dt} for the composite estimator. A characteristic common to most approaches is that the weights have considerable variation across geographical domains. This is a result of the fact that the sizes of the domain samples used for the direct estimator often vary dramatically when samples are designed to make estimates for a higher level of geographic aggregation. The three program applications described in Chapters 7 through 9 are distinguished by different indirect component estimators and different approaches to estimating the composite estimator weight.

It should be noted that not all indirect estimators are linear. For examples of nonlinear indirect estimators see MacGibbon and Tomberlin (1989) and Malec, Sedransk, and Tompkins (1993). This latter, nonlinear indirect estimator is being considered for use in conjunction with the National Health Interview Survey and is discussed in Chapter 8.

1.4 Program Summaries

The programs described in this report were initiated in response to a variety of needs and directives. Several are a direct result of legislative requirements to allocate federal funds. Others were created in response to state needs for data and/or to standardize estimation methods across states. Others are viewed as research programs that periodically publish estimates when an improved methodology has been developed. Table 2 below allows a comparison of summary information on the programs described in Chapters 2 through 9 in this report.

Table 2. Selected Characteristics of U.S. Federal Programs that Use Indirect Estimators to Publish Estimates

Ch.	Estimator	Variables	Domain	Frequency
2	Domain indirect synthetic	Infant and maternal health characteristics	States	Periodically
3	Domain indirect regression (ratio)	Personal income, annual income, gross product	States and counties	Annually (Quarterly)
4	Domain and time indirect regression	Postcensal populations	Counties	Annually
5	Time indirect regression	Employment and unemployment	States	Monthly
6	Domain indirect regression	Cotton, rice, and soybean acreage	Counties	Annually
7	Time indirect composite	Livestock inventories, crop production and acreage	Counties	Annually
8	Domain indirect composite	Disabilities, hospital utilization, physician and dental visits	States	Periodically
9	Domain indirect composite	Median income for 4-person families	States	Annually

As noted earlier, a variety of indirect estimators are used to produce estimates. Synthetic, regression, and composite estimators that borrow strength over domains, over time, or over both domain and time are found among these programs. The estimation procedures for six of the programs are based on data from sample surveys. There is no sampling involved in the procedures used in the two programs that produce estimates of personal income and postcensal populations. In some instances, a program produces estimates for a single variable; in other instances, estimates are produced for numerous variables. States and counties are the only domains for which indirect estimates are presently published. Four of the programs publish estimates for states, three for counties, and one for both states and counties.

There is considerable variability in the frequency with which different estimates are published. Given the differing demands on Federal statistical agencies, it is not surprising that considerable variation is seen in the programs described in this report.

REFERENCES

Brackstone, G. J. (1987), "Small Area Data: Policy Issues and Technical Challenges," in *Small Area Statistics*, New York: John Wiley and Sons.

Dalenius, T. (1987), "Panel Discussion" in *Small Area Statistics*, New York: John Wiley and Sons.

Ericksen, E.P. (1974), "A Regression Method for Estimation Population Changes for Local Areas," *Journal of the American Statistical Association*, 69, 867-875.

Fay, R.E. and Herriott, R.A. (1979), "Estimates of Income for Small Places: An Application of James-Stein Procedures to Census Data," *Journal of the American Statistical Association*, 74, 269-277.

Ghosh, M. (1992), "Constrained Bayes Estimation With Applications," *Journal of the American Statistical Association*, 87, 533-540.

Gonzalez, M.E. (1973), "Use and Evaluation of Synthetic Estimates," *Proceedings of the Social Statistics Section*, American Statistical Association, 33-36.

Gonzalez, M.E. (1979), "Case Studies on the Use and Accuracy of Synthetic Estimates: Unemployment and Housing Applications" in *Synthetic Estimates for Small Areas* (National Institute on Drug Abuse, Research Monograph 24), Washington, D.C.: U.S. Government Printing Office.

Gonzalez, M.E. and Hoza, C. (1978), "Small-Area Estimation with Application to Unemployment and Housing Estimates," *Journal of the American Statistical Association*, 73, 7-15.

Gonzalez, M.E. and Waksberg, J (1973), "Estimation of the Error of Synthetic Estimates," paper presented at the first meeting of the International Association of Survey Statisticians, Vienna, Austria, 18-25 August, 1973.

Heeringa, S.G. (1981), "Small Area Estimation Prospects for the Survey of Income and Program Participation," *Proceedings of the Section on Survey Research Methods,* American Statistical Association, 133-138.

Holt, D., Smith, T.M.F., and Tomberlin, T.J. (1979), "A Model-Based Approach to Estimation for Small Subgroups of a Population," *Journal of the American Statistical Association,* 74, 405-410.

Kalton, G. (1987), "Panel Discussion" in *Small Area Statistics,* New York: John Wiley and Sons.

Laake, P. (1979), "A Prediction Approach to Subdomain Estimation in Finite Populations," *Journal of the American Statistical Association,* 74, 355-358.

Lahiri, P. (1990), "Adjusted Bayes and Empirical Bayes Estimation in Finite Population Sampling," *Sankhya B,* 52, 50-66.

MacGibbon, B. and Tomberlin, T.J. (1989), "Small Area Estimation of Proportions Via Empirical Bayes Techniques," *Survey Methodology,* 15-2, 237-252.

Malec, D., Sedransk, J., and Tompkins, L. (1993), "Bayesian Predictive Inference for Small Areas for Binary Variables in the National Health Interview Survey." In *Case Studies in Bayesian Statistics,* eds., Gatsonis, Hodges, Kass and Singpurwalla. New York: Springer Verlag.

National Center for Health Statistics (1968), *Synthetic State Estimates of Disability* (PHS Publication No. 1759), Washington, D.C.: U.S. Government Printing Office.

National Institute on Drug Abuse (1979), *Synthetic Estimates for Small Areas* (NIDA Research Monograph 24), Washington, D.C.: U.S. Government Printing Office.

Prasad, N.G.N. and Rao, J.N.K. (1990), "The Estimation of the Mean Squared Error of Small-Area Estimators," *Journal of the American Statistical Association,* 85, 163-171.

Purcell, N.J. and Kish, L. (1979), "Estimation for Small Domains," *Biometrics,* 35, 365-384.

Räbäck, G. and Särndal, C.E. (1982), "Variance Reduction and Unbiasedness for Small Area Estimators," *Proceedings of the Social Statistics Section,* American Statistical Association, 541-544.

Royall, R.A. (1973), "Discussion of papers by Gonzalez and Ericksen," *Proceedings of the Social Statistics Section*, American Statistical Association, 42-43.

Royall, R.A. (1979), "Prediction Models in Small Area Estimation," in *Synthetic Estimates for Small Areas* (National Institute on Drug Abuse, Research Monograph 24), Washington, D.C.: U.S. Government Printing Office.

Särndal, C.E. (1984), "Design-Consistent versus Model-Dependent Estimation for Small Domains," *Journal of the American Statistical Association*, 79, 624-631.

Särndal, C.E. and Hidiroglou, M.A. (1989), "Small Domain Estimation: A Conditional Analysis," *Journal of the American Statistical Association*, 84, 266-275.

Schaible, W.L. (1993), "Indirect Estimators: Definition, Characteristics, and Recommendations," *Proceedings of the Section on Survey Research Methods*, American Statistical Association, 1-10.

Schaible, W.L., Brock, D.B., and Schnack, G.A. (1977), "An Empirical Comparison of the Simple Inflation, Synthetic and Composite Estimators for Small Area Statistics," *Proceedings of the Social Statistics Section*, American Statistical Association, 1017-1021.

Schaible, W.L., Brock, D.B., Casady, R.J., and Schnack, G.A. (1979), *Small Area Estimation: An Empirical Comparison of Conventional and Synthetic Estimators for States*, (PHS Publication No. 80-1356), Washington, D.C.: U.S. Government Printing Office.

Spjøvoll, E. and Thomsen, I. (1987), "Application of Some Empirical Bayes Methods to Small Area Statistics," *Proceedings of the International Statistical Institute*, Vol. 2, 435-449.

U.S. Office of Management and Budget (1993), "Indirect Estimators in Federal Programs," *Statistical Policy Working Paper 21*, National Technical Information Service, (NTIS Document Sales, PB93-209294) Springfield, Virginia.

Woodruff, R.S., (1966), "Use of a Regression Technique to Produce Area Breakdowns of the Monthly National Estimates of Retail Trade," *Journal of the American Statistical Association*, 79, 496-504.

CHAPTER 2

Synthetic Estimation in Followback Surveys at The National Center for Health Statistics

Joe Fred Gonzalez, Jr., Paul J. Placek, and Chester Scott
National Center for Health Statistics

2.1 Introduction and Program History

The National Center for Health Statistics (NCHS) through its vital registration system collects and publishes data on vital events (births and deaths) for the United States (NCHS 1989). NCHS produces national, State, county, and smaller area vital statistics for sociodemographic and health characteristics which are available from birth and death certificates. The Division of Vital Statistics of NCHS produces annual summary tables for the United States showing trends in period and cohort fertility measures and characteristics of live births. Also, NCHS produces detailed tabulations by place of residence and occurrence for each State, county, and city with a population of 10,000 or more; by race and place of delivery and place of residence for population-size groups in metropolitan and nonmetropolitan counties within each State by race, attendant and place of delivery, and birth weight. These statistics are based on a complete count of vital records.

In addition to the limited vital statistics tabulations which are produced annually, there has always been a continuing need for more detailed national and State level estimates of health status, health services, and health care utilization related to vital events.

Because vital records (birth and death certificates) serve both legal and statistical purposes, they provide limited social, demographic, health, and medical information. Each vital record is a one page document with extremely limited information. The data from these vital records can be augmented, however, through periodic "followback" surveys. These surveys are referred to as "followback" because they obtain additional information from sources named on the vital record. A followback survey is a cost effective means of obtaining supplementary information for a sample of vital events. From the sample it is possible to make national estimates of vital events according to characteristics not otherwise available. Examples of supplementary information which may be needed by health researchers, health program planners, and health policy makers are: mother's smoking habits before and during pregnancy; complications of pregnancy; drug or surgical procedure to induce or maintain labor; amniocentesis during pregnancy; electronic fetal monitoring; respiratory distress syndrome; infant jaundiced; medical x-ray use; birth injuries; and, congenital anomalies. Periodic followback surveys respond to the changing data needs of the public health community without requiring changes in the vital record forms.

The specific NCHS followback surveys that will be discussed in this chapter are the 1980 National Natality Survey (NNS) and the 1980 National Fetal Mortality Survey (NFMS) (NCHS 1986). In order to produce State estimates for certain health characteristics not available on the vital records, synthetic estimation (NCHS 1984a, 1984b) was applied to national data from the 1980 NNS and 1980 NFMS. In addition to the usual appeal of using synthetic estimation over direct estimation, especially when small sample sizes are concerned, synthetic State estimates were compared to direct State estimates as well as the "true" values for a limited number of variables from State vital statistics via fetal death records and birth and death certificates.

2.2 Program Description, Policies and Practices

The 1980 NNS is based on a probability sample of 9,941 from a universe of 3,612,258 live births that occurred in the United States during 1980. The NNS sample included a four-fold oversampling of low birth weight infants. The live birth certificate represents the basic source of information. Based on information from the sample birth certificates, eight page Mother's questionnaires were mailed to mothers who were married. These mothers were asked to provide information on prenatal health practices, prenatal care, previous pregnancies, and social and demographic characteristics of themselves and their husbands. Each mother was also asked to sign a consent statement authorizing NCHS to obtain supplemental information from her medical records. If the mother did not respond after two questionnaires were sent by mail, a telephone interviewer attempted to complete an abbreviated questionnaire and to obtain a consent statement. To ensure their privacy, unmarried mothers were not contacted. As a result of sending the Mother's questionnaire only to married mothers, the 1980 NNS population of inference for data collected through the Mother's questionnaire was 2,944,580 live births.

Regardless of the mother's marital status, questionnaires were mailed to the hospital's and to the attendants at delivery (for example, physicians or nurse-midwives) named on the birth certificates. A questionnaire was sent to the hospital for each sample birth that occurred either in a hospital or en route to a hospital. If the mother signed a consent statement authorizing NCHS to obtain supplemental medical information, a copy was included with the questionnaire. The focus of the hospital questionnaire was on characteristics of labor and delivery, health characteristics of the mother and infant, information on prenatal care visits, and information on radiation examinations and treatments received by the mother during the 12 months before delivery of the sample birth. For the hospital component of the 1980 NNS, the population of inference was 3,580,700 live births.

The 1980 NNS is composed of information from birth certificates and information from questionnaires sent to married mothers, hospitals, attendants at delivery, and providers of radiation examinations and treatments. The survey represents an extensive source of information concerning specific maternal and child health conditions and obstetric practices for live births in the United States. The 1980 NNS response rates were 79.5 percent for mothers, 76.1 percent for hospitals, and 61.6 percent for physicians.

The 1980 NFMS is based on a probability sample of 6,386 fetal deaths (out of a universe of 19,202 fetal deaths) with gestation of 28 weeks or more, or delivery weight of 1,000 grams or more, that occurred in the United States during 1980. The report of fetal death represent the basic source of information in this survey. Married mothers, hospitals, attendants at delivery, and providers of radiation examinations and treatments were surveyed under the same conditions as those described for the 1980 NNS. The 1980 NFMS populations of inference for all fetal deaths, fetal deaths in hospitals, and fetal deaths to married mothers were 19,202, 18,930, and 14,790, respectively. The same questionnaires were used for both surveys. Although some questions pertained only to live births and other pertained only to fetal deaths, instructions to skip inappropriate questions were included in the questionnaires. The sampling design for the NFMS was developed so that the NFMS would be large enough to permit comparisons between live births in the NNS and fetal deaths in the NFMS. The 1980 NFMS response rates were 74.5 percent for mothers, 74.0 percent for hospitals, and 55.0 percent for physicians.

Table 1 presents the 1980 NNS and NFMS distribution of sample cases of live births and fetal deaths by State of occurrence. As shown in Table 1, it may be possible to produce direct State level estimates of certain health characteristics for some of the larger States. However, the sample sizes for most States are generally too small to produce reliable direct State estimates. This was the main justification for exploring synthetic State estimation as an alternative for producing State level estimates.

Table 1. Number of Sample Cases by State of Occurrence: 1980 National Natality and National Fetal Mortality Surveys

State	Live Births	Fetal Deaths	State	Live Births	Fetal Deaths
Total	9,941	6,386	Missouri	233	160
Alabama	178	172	Montana	44	23
Alaska	25	19	Nebraska	74	60
Arizona	120	80	Nevada	35	29
Arkansas	105	96	New Hampshire	37	18
California	1,077	615	New Jersey	207	99
Colorado	144	100	New Mexico	50	30
Connecticut	106	64	New York	705	481
Delaware	25	15	North Carolina	230	216
District of Columbia	61	44	North Dakota	36	17
Florida	355	268	Ohio	493	322
Georgia	284	250	Oklahoma	147	106
Hawaii	44	34	Oregon	112	74
Idaho	46	24	Pennsylvania	468	341
Illinois	537	413	Rhode Island	36	27
Indiana	248	141	South Carolina	142	149
Iowa	141	86	South Dakota	37	22
Kansas	110	84	Tennessee	211	185
Kentucky	164	135	Texas	696	413
Louisiana	225	166	Utah	121	75
Maine	42	30	Vermont	21	9
Maryland	150	86	Virginia	216	41
Massachusetts	191	108	Washington	164	85
Michigan	411	...[1]	West Virginia	84	66
Minnesota	190	101	Wisconsin	204	55
Mississippi	131	135	Wyoming	28	17

[1] Participation precluded by State law.

2.3 Estimator Documentation

The underlying rationale for synthetic estimation is that the distribution of a health characteristic is highly related to the demographic composition of the population (NCHS 1984a). It is assumed that differences in the prevalence of the characteristics between two areas are due primarily to differences in demographic composition (e.g. age, race, sex, etc.). That is, it is assumed that a particular measure would be the same in two States that had the same population composition with respect to certain demographic variables. This rationale was used to select the demographic variables

that were deemed to be the most appropriate and relevant to the 1980 NNS and
NFMS in order to produce Synthetic State estimates.

The following is the basic estimator that was used to produce Synthetic State
estimates of proportions for certain health variables from the 1980 National Natality
Survey (NNS) and the 1980 National Fetal Mortality Survey (NFMS).

$$\hat{P}_{(SYN),s} = \frac{\sum_i \sum_j \sum_k \hat{P}_{ijk} N_{ijks}}{\sum_i \sum_j \sum_k N_{ijks}}$$

where

i	$=$	mother's race (white and all other),
j	$=$	mother's age group (6 groups),
k	$=$	live birth order (1, 1-2, 1-3, 2, 2+, 3, 3+, 4+),
$\hat{P}_{ijk}$	$=$	estimated national NNS or NFMS proportion of births (or fetal deaths) within a certain ijk subdomain having a certain health characteristic,
N_{ijks}	$=$	the number of births (or deaths) within a certain ijk subdomain within the sth State (derived from State Vital registration data).

Table 2 gives an illustration of the computation of the synthetic State estimate of the
percent jaundiced infants in Pennsylvania in 1980. The stub of Table 2 shows the 25
demographic cells (race, age of mother, and live-birth order groups) that were used
to produce the Synthetic State estimates. Column (1) shows the national (based on
the 1980 NNS) estimates of percent of live births that were jaundiced in each of the
respective 25 demographic cells. Column (2) shows the cells in Pennsylvania.
Column (3), the estimated number of jaundiced live births in Pennsylvania, is
computed by taking the product of entries in columns (1) and (2) within each of the
25 respective cells. Finally, the Synthetic State estimate is found by taking the ratio
of the sum of column (3) to the sum of column (2).

21

Table 2. Illustration of Computation of Synthetic State Estimate of Percent Jaundiced Live Births in Pennsylvania, 1980

Race, age, live-birth order groups	Percent jaundiced, U.S., NNS	Hospital births, Pennslyvania	Estimated number of jaundiced infants, Pennsylvania
	(1)	(2)	(3)=(1)x(2)
White, under 20, 1	.205	13,484	2,764
White, under 20, 2+	.265	2,898	768
White, 20-24, 1	.228	23,739	5,412
White, 20-24, 2	.200	15,005	3,001
White, 20-24, 3+	.189	5,356	1,012
White, 25-29, 1	.231	17,664	4,080
White, 25-29, 2	.222	17,877	3,969
White, 25-29, 3	.232	7,959	1,846
White, 25-29, 4+	.139	2,921	406
White, 30-34, 1	.223	4,889	1,090
White, 30-34, 2	.199	8,115	1,615
White, 30-34, 3	.263	5,768	1,517
White, 30-34, 4+	.219	3,628	795
White, 35+, 1-3	.284	3,445	978
White, 35+, 4+	.188	2,451	461
All other, under 20, 1	.191	4,208	804
All other, under 20, 2+	.114	1,285	146
All other, 20-24, 1	.187	3,174	594
All other, 20-24, 2	.149	2,704	403
All other, 20-24, 3+	.158	1,779	281
All other, 25-29, 1	.215	1,390	299
All other, 25-29, 2	.176	1,674	295
All other, 25-29, 3+	.185	1,999	370
All other, 30+, 1-2	.257	1,535	394
All other, 30+, 3+	.273	1,852	506
Total		156,799	33,806

Synthetic State estimate= 33,806 ÷ 156,799 = 21.6%

Since there were three different populations of inference (all vital events, vital events to married mothers, and vital events in hospitals) for each of the 1980 NNS and NFMS, appropriate State aggregates of vital events were incorporated into the calculation of corresponding synthetic State estimates (NCHS 1984a, 1984b).

2.4 Evaluation Practices

The following is a description of some of the tabulations that were produced. Table 3 gives Synthetic State estimates of 11 health characteristics of mothers and infants for five selected States. A complete listing of all 57 NNS/NFMS health variables for which Synthetic State estimates were produced can be found in Tables 2-8 in (NCHS 1984a, 1984b).

Table 3. Synthetic State estimates of selected characteristics of mothers and infants for 5 states: 1980 National Natality Survey

Characteristics	Pennsylvania	Indiana	Tenn.	Kansas	Montana
All births			Percent		
Ultrasound	33.7	33.2	33.4	33.3	33.3
Hospital births					
Amniocentesis	4.4	4.1	4.2	4.1	4.2
Electronic fetal monitoring	47.7	47.3	48.0	47.5	47.3
Induction or maintenance of labor	43.2	43.0	43.2	43.1	43.0
Cesarean delivery	17.3	16.7	17.0	16.8	16.8
Postpartum sterilization	10.2	10.3	9.8	10.1	10.5
Respiratory distress	3.9	4.0	4.1	3.9	3.9
Infant jaundiced	21.6	21.4	21.1	21.5	21.5
Married mothers					
Smoked during pregnancy	26.1	27.6	27.3	27.2	26.7
Drank during pregnancy	4.9	4.6	4.6	4.6	4.7
Worked during pregnancy	62.5	60.1	61.6	61.0	61.1

Table 4 shows a comparison of the "true" State values (p) with the NNS direct State estimates ($\hat{p}$) of percent low birth weight, percent late or no prenatal care, and percent low 1-minute Apgar score for five selected States in 1980. Table 5 compares the "true" State values (p) with the NNS synthetic State estimates ($\hat{p}_{(SYN),s}$) for the same health variables in Table 4. The five States that are shown in Tables 3 through 5 were selected for comparison because they covered a wide range in the annual number of births, from about 15,000 to nearly 160,000. The "true" values in Tables 4 and 5 were produced from the entire birth certificate and fetal report state files which are produced through the State vital registration system. Unfortunately as mentioned earlier, there are only a few "true" values that can be produced since the data originates from the vital records which have very limited information.

The synthetic State estimates are subject to sampling error because they are based on corresponding national estimates derived from the 1980 NNS and NFMS by race, maternal age, and live-birth order group. Therefore, the standard errors of the synthetic State estimates are relatively small because they are based on the standard errors of the national estimates. The standard errors for the NNS and NMFS were estimated by a balanced-repeated-replicated procedure using 20 replicate half samples. This procedure estimates the standard errors for survey estimates through the observation of the variability of estimates based on replicate half samples of the total sample. This variance estimation procedure was developed and described by McCarthy (NCHS 1966, 1969).

Although the synthetic State estimate has a relatively small standard error, it is subject to bias. The bias is a measure of the extent to which the national maternal age, race, live-birth order specific estimates differ from the true values for a given State. The closer the demographic variables used in the synthetic estimate come to accounting for all the interstate variation in a particular health characteristic, the smaller the bias will be. Unfortunately, the bias cannot be computed without knowing the true values. However, through the U.S Vital Registration System, true State values for vital events (collected through birth and death certificates) are known for a limited number of available sociodemographic and health characteristics. Therefore, we can compare certain synthetic estimates with their corresponding true values. This yields a degree of confidence for the synthetic estimates of similar characteristics which cannot be checked against the true values from State vital statistics. Thus, the evaluation of this study only provides an indicator of the quality of the synthetic State estimates.

Table 4. Comparison of Vital registration ("true" value) data percent and NNS direct estimates of percent low birth weight, percent late or no prenatal care, and percent low 1-minute Apgar score: Selected States, 1980 National Natality Survey

Characteristic and State	Vital registration "True" Percent	NNS, direct estimate			Direct Estimate	
		Percent	Standard error	Relative standard error	MSE	RRMSE
Low birth						
United States	6.8	6.9	0.2	0.03	.04	
Pennsylvania	6.5	6.6	1.0	0.16	1.00	.15
Indiana	6.3	6.8	1.4	0.21	1.96	.22
Tennessee	8.0	8.5	1.8	0.22	3.24	.23
Kansas	5.8	6.8	2.1	0.31	4.41	.36
Montana	5.6	9.2	4.0	0.43	16.00	.71
Prenatal care[1]						
United States	5.1	4.7	0.2	0.04	.04	
Pennsylvania	3.9	4.3	0.8	0.20	.64	.21
Indiana	3.8	2.0	0.8	0.39	.64	.21
Tennessee	5.4	4.7	1.4	0.29	1.96	.26
Kansas	3.4	2.1	1.2	0.57	1.44	.35
Montana	3.7	3.0	2.3	0.79	5.29	.62
Apgar score[2]						
United States	9.4[3]	9.6	0.3	0.03	.09	
Pennsylvania	7.9	7.7	1.1	0.14	1.21	.14
Indiana	10.9	9.5	1.7	0.17	2.89	.16
Tennessee	9.6	7.3	1.7	0.23	2.89	.18
Kansas	11.1	12.3	2.8	0.22	7.84	.25
Montana	11.6	12.9	4.6	0.36	21.16	.40

[1] Late (beginning in third trimester) or no prenatal care
[2] Low (less than 7) 1-minute Apgar score
[3] Total of 44 reporting States

Table 5. Comparison of Vital registration ("true" value) data percent and NNS synthetic estimates of percent low birth weight, percent late or no prenatal care, and percent low 1-minute Apgar score: Selected States, 1980 National Natality Survey

Characteristic and State	Vital registration "True" Percent	NNS, synthetic estimate Percent	NNS, synthetic estimate Standard error	NNS, synthetic estimate Bias	Synthetic Estimate MSE	Synthetic Estimate RRMSE
Low birth						
United States	6.8	...	...	...		
Pennsylvania	6.5	6.5	0.2	0.0	.00	.00
Indiana	6.3	6.5	0.2	0.2	.04	.03
Tennessee	8.0	7.2	0.2	-0.8	.64	.10
Kansas	5.8	6.4	0.2	0.6	.36	.10
Montana	5.6	6.3	0.2	0.7	.49	.13
Prenatal care[1]						
United States	5.1	...	...	...		
Pennsylvania	3.9	4.3	0.2	0.4	.16	.10
Indiana	3.8	4.7	0.2	0.9	.81	.24
Tennessee	5.4	5.0	0.2	-0.4	.16	.07
Kansas	3.4	4.5	0.2	1.1	1.21	.32
Montana	3.7	4.3	0.2	0.6	.36	.16
Apgar score[2]						
United States	9.4[3]	...	...	...		
Pennsylvania	7.9	9.4	0.3	1.5	2.25	.19
Indiana	10.9	9.4	0.3	-1.5	2.25	.14
Tennessee	9.6	9.7	0.3	0.1	0.01	.01
Kansas	11.1	9.4	0.3	-1.7	2.89	.15
Montana	11.6	9.4	0.3	-2.2	4.84	.19

[1] Late (beginning in third trimester) or no prenatal care
[2] Low (less than 7) 1-minute Apgar score
[3] Total of 44 reporting States

The second to the last column of Tables 4 and 5 show the estimated mean square error (MSE) of the NNS direct estimates and the NNS synthetic estimates, respectively. The MSE of an estimate x is the variance of x plus the square of the bias of x, i.e.,

$$MSE(x) = VAR(x) + [BIAS(x)]^2.$$

Since the NNS direct State estimates are unbiased, the MSE of the direct State estimates are equal to the variance of the direct State estimates. The unbiased estimator for the MSE of the synthetic estimates used for Table 5 was $(\hat{p}_{(SYN),s} - p)^2$. As shown in the second to the last column of Tables 4 and 5, the synthetic estimator performed better than the direct estimator with respect to a smaller MSE, with only two exceptions. In general, the synthetic estimator performed better than the direct estimator for the smaller States where the number of NNS sample cases was small. In order to compare the precision of the synthetic State estimates with the the precision of the direct State estimates, the relative root mean square error (RRMSE) was calculated for both the synthetic and direct estimates. The RRMSE was calculated by taking the ratio of the square root of the MSE to the "true percent." The NCHS standard for reliability of estimates for followback surveys has been a maximum relative standard error (RSE) of 0.25. As shown in the last column of Table 4 and Table 5, the RRMSE for the synthetic estimator was smaller (with two exceptions) than the RRMSE for the direct estimator and shows that most synthetic estmates would meet the NCHS criterion for reliabilty.

2.5 Current Problems And Activities

Work is currently underway at NCHS to produce synthetic State estimates from the 1988 National Maternal and Infant Health Survey (NMIHS) which is very similar to its predecessor the 1980 NNS and NFMS. In the NMIHS 9,953 out of a universe of 3,898,922 live-birth certificates are linked with mothers' responses on 35-page questionnaires about the mothers' prenatal health behavior, maternal health, the birth experience, and infant health. The 1988 NMIHS live birth estimates will be used to produce synthetic State estimates by infant's race, birth weight, and maternal age and marital status.

REFERENCES *

National Center for Health Statistics: Vital Statistics of the United States, 1987 Vol. 1, Natality, DHHS Pub. No. (PHS) 89-1100. Public Health Service, Washington. U.S. Government Printing Office, 1989.

National Center for Health Statistics, K.G. Keppel, R.L. Heuser, P.J. Placek, et al.: Methods and Response Characteristics, 1980 National Natality and Fetal Mortality Surveys. Vital and Health Statistics, Series 2, No. 100. DHHS Pub No. (PHS) 86-1374. Public Health Service, Washington. U.s. Government Printing Office, Sept. 1986.

National Center for Health Statistics: State Uses of Followback Survey Data, R.L. Heuser, K.G. Keppel, C.A. Witt, and P.J. Placek, Presented at the Annual Meeting of the Association for Vital Records and Health Statistics, July 9-12, 1984, Niagara Falls, NY.

National Center for Health Statistics: R.L. Heuser, K.G. Keppel, C.A. Witt, and P.J. Placek, Synthetic Estimation Applications form the 1980 National Natality Survey (NNS) and the 1980 National Fetal Mortality Survey (NFMS), Presented at the NCHS Data Use Conference on Small Area Statistics, August 29-31, 1984, Snowbird, Utah.

National Center for Health Statistics, P.J. McCarthy: Replication: An Approach to the Analysis of Data From Complex Surveys. Vital and Health Statistics, Series 2, No. 14, PHS Pub No. 1000. Public Health Service. Washington, U.S. Government Printing Office, April 1966.

National Center for Health Statistics, P.J. McCarthy: Pseudoreplication: Further Evaluation and Application of the Balanced Half-Sample Technique. Vital and Health Statistics. Series 2, No. 31. DHEW Pub No. (HSM) 73-120. Health Services and Metal Health Administration. Washington. U.S. Government Printing Office, Jan. 1969.

* Chapter 8 (authored by Donald Malec) of this report contains several references on small area estimation as applied to the National Health Interview Survey of the National Center for Health Statistics.

CHAPTER 3

State, Metropolitan Area, and County Income Estimation

Wallace Bailey, Linnea Hazen, and Daniel Zabronsky
Bureau of Economic Analysis

3.1 Introduction and Program History

3.1.1 Program Description

The Bureau of Economic Analysis (BEA) maintains a program of State and local area (county and metropolitan area) economic measurement that centers on the personal income measure. This program originated in 1939 when estimates of income payments to individuals by State were first published. At the national level, personal income is the principal income measure in the personal income and outlay account, one of the five accounts that compose the national income and product accounts. The State and local area personal income estimates are derived by disaggregating the detailed components of the national personal income estimates to States and counties. Estimates for all other geographic areas are made by aggregating either the State or county estimates in the appropriate combinations. This building block approach permits estimates for areas whose boundaries change over time, such as metropolitan areas, to be presented on a consistent geographic definition for all years.

3.1.2 Uses of the State and Local Area Income Estimates

BEA's State and local area income estimates are widely used in the public and private sector to measure and track levels and types of incomes received by persons living or working in an area. They provide a framework for the analysis of each area's economy and serve as a basis for decision making in both the public and private sectors. Personal income is among the measures used in evaluating the socioeconomic impact of public- and private-sector initiatives; for example, it is widely used in preparing the environmental impact statements required by the National Environmental Policy Act of 1969.

One of the first uses made of State personal income estimates (or a derivative) was as a variable in formulas for allocating Federal funds to States. The most often used derivative is per capita personal income, which is computed using the Census Bureau's estimates of total population; these population estimates are described by Long in Chapter 4 of this report. At present, BEA's State personal income estimates are used by the Federal Government to allocate over $92 billion annually for various Federal domestic programs, including the medical assistance (medicaid) program, and the aid to families with dependent children program. Table 3.1 highlights the major Federal Government programs which use BEA personal income estimates in allocation formulas for Federal domestic assistance funds.

Federal agencies also use the components of personal income in econometric models, such as those used to project energy and water use. The U.S. Forest Service is using these estimates to identify resource dependent rural areas and to allocate funds for their economic diversification as required by the National Forest-Dependent Rural Communities Economic Diversification Act of 1990.

The U.S. Census Bureau uses the BEA estimates of State per capita personal income as the key predictor variable in its estimates of mean annual income for 4-person families by State. These estimates are described by Fay, Nelson, and Litow in Chapter 9 of this report.

During the past decade, State governments have substantially increased their use of the State personal income estimates. The estimates are used in the measurement of economic bases and in models developed for planning for such things as public utilities and services. They are also used to project tax revenues. In recent years, legislation that limits a State's expenditures or tax authority by the level of, or changes in, State personal income or to one of its components has been enacted in 16 States. These 16 States account for nearly one-half of the U.S. population. Some of these States used BEA's annual State personal income estimates; the others use fiscal year estimates derived from BEA's quarterly State personal income estimates (ACIR, 1990).

State governments also use the local area estimates to measure the economic base of State planning areas. University schools of business and economics, often working under contract for State and local governments, use the BEA local area estimates for theoretical and applied economic research.

Businesses use the estimates to evaluate markets for new or established products and to determine areas for the location, expansion, and contraction of their activities. Trade associations and labor organizations use them for product and labor market analyses.

Table 3.1. Programs Using BEA Personal Income Estimates in Allocation Formulas for Federal Domestic Assistance Funds, Fiscal Year 1992

Program Number	Program Name	FY 1992 Obligations (Millions of $)
17.235	Senior Community Service Employment Program	395.2
84.126	Rehabilitation Services	1,783.5
84.154	Public Library Construction and Technology Enhancement	29.8
93.020	Family Support Payments to States (AFDC)	13,814.9
93.138	Protection and Advocacy for Mentally Ill Individuals	19.1
93.630	Developmental Disabilities Basic Support and Advocacy	90.2
93.645	Child-Welfare Services--State Grants	273.9
93.658	Foster Care--Title IV-E	2,342.1
93.659	Adoption Assistance	201.9
93.778	Medical Assistance Program (Medicaid; Title XIX)	72,502.7
93.779	Health Care Financing Research	78.4
93.992	Alcohol & Drug Abuse & Mental Health Services	1,292.0
TOTAL		92,823.7

Source: Office of Management and Budget and U. S. General Services Administration (1992), 1992 Catalog of Federal Domestic Assistance, Washington, DC: U.S. Government Printing Office. For information about grant formulas, see U. S. General Services Administration (1992), 1992 Formula Report to the Congress, Washington, DC: U.S. Government Printing Office.

3.1.2 A History of BEA's Regional Income Estimates

In the mid-1930's, BEA's predecessor began work on the estimation of regional income as part of the effort to explain the processes and structure of the Nation's economy. As a result of its work, it produced a report that showed State estimates of total "income payments to individuals" in May 1939 (Nathan and Martin, 1939). These income payments were defined as the sum of (1) wages and salaries, (2) other labor income and relief, (3) entrepreneurial withdrawals, and (4) dividends, interest, and net rents and royalties.

In 1942, the State estimates of wages and salaries and entrepreneurial income were expanded to include a further breakdown by broad industry group--agriculture, other commodity-producing industries, distribution, services, and government. The industry breakdown was for 1939, when the availability of census information on payrolls and the employed labor force by industry and by State made possible more reliable estimates than for prior years (Creamer and Merwin, 1942). The estimates for most nonagricultural industries and for the military services were based on reports in which establishments, not employees, were classified by State and in which the State of residence of the employees was not indicated; therefore, the estimates for these industries were on a "place-of-work" (where-earned) basis. No systematic adjustment was made in the total income payments series to convert the estimates to a "place-of-residence" (where-received) basis. However, using the limited information that was available, residence adjustments were made for a few States for the per capita series.

During the later 1940's and early 1950's, BEA continued to work on improving these estimates by seeking additional source data and by improving the estimating methods that were used. The industrial detail of the wage and salary estimates was expanded to include each Standard Industrial Classification (SIC) division and additional detail for some SIC divisions. As one result of the major reworking and expansion of the national income and product accounts, BEA developed State personal income--a measure of income that is more comprehensive than State income payments.

During the 1960's and 1970's, BEA continued its work to provide more information about regional economies. Annual State estimates of disposable personal income were published in the April 1965 Survey of Current Business (Survey), and the first set of quarterly estimates of State personal income was published in the December 1966 Survey. Estimates of personal income for metropolitan areas were published in the May 1967 Survey, for nonmetropolitan counties in the May 1974 Survey, and for metropolitan counties in the April 1975 Survey. In the late 1970's, BEA introduced annual estimates of employment for States, metropolitan areas, and counties.

Refinement of the residence adjustment procedures and a fuller presentation of industrial detail for earnings--the term introduced to cover wages and salaries plus other labor income plus proprietors' income--emerged in the estimates published in 1974. The residence adjustment procedures had been extended to all States in 1966, but the residence adjustment estimates (i.e., the net flows of interstate commuters' earnings), along with earnings by industry on a place-of-work basis, were not published explicitly until 1974.

3.2 The Regional Economic Measurement Program

3.2.1 Estimating Schedule for State and Local Area Personal Income Series

The annual estimates of State personal income for a given year are subject to successive refinement. Preliminary estimates, based on the current quarterly series, are published each April, 4 months after the close of the reference year, in the Survey. The following August, more reliable annual estimates are published. These estimates are developed independently of the quarterly series and are prepared in greater component detail, primarily from Federal and State government administrative records. The annual estimates published in August are subsequently refined to incorporate newly available information used to prepare the local area estimates for the same year. These revised State estimates, together with the local area estimates, are published the following April. The annual estimates emerging from this three-step process are subject to further revision for several succeeding years (the State estimates in April and August and the local area estimates in April), as additional data become available. For example, the 1992 State estimates that were first released in April 1993 will be revised in August 1993 and in April and August of subsequent years; the 1991 local area estimates that were first released in April 1993 were revised in April 1994, and will be further revised in April of subsequent years. The routine revisions of the State estimates for a given year are normally completed with the fourth April publication, and the local area estimates, with the third April publication. After that, the estimates will be changed only to incorporate a comprehensive revision of the National Income and Product Accounts--which takes place approximately every 5 years--or to make important improvements to the estimates through the use of additional or more current State and local area data.

Quarterly estimates of State personal income, which are available approximately 4 months after the close of the reference quarter, are published regularly in the January, April, July, and October issues of the Survey. In October and again the following April, the quarterly series for the 3 previous years is revised for consistency with the revised annual estimates. In January and July, at least the quarter immediately preceding the current quarter is revised.

3.2.2 Availability of State and Local Area Estimates

The State and local area personal income and employment estimates are available through the Regional Economic Information System (REIS), which operates an information retrieval service that provides a variety of standard and specialized analytic tabulations for States, counties and specified combinations of counties. Standard tabulations include personal income by type and earnings by industry, employment by industry, transfer payments by program, and major categories of farm gross income and expenses. These tabulations are available from REIS in magnetic tape, computer printout, microcomputer diskette and CD-ROM; some of the tabulations are also available electronically on the Department of Commerce's Economic Bulletin Board, available through the National Technical Information Service. In addition, summary tabulations of the State and local estimates are published regularly in BEA's major publication, the Survey. An extensive set of State-level historical estimates is available (BEA, 1989).

BEA also makes its regional estimates available through the BEA User Group, members of which include State agencies, universities, and Census Bureau Primary State Data Centers. BEA provides its estimates of income and employment for States, metropolitan areas, and counties to these organizations with the understanding that they will make the estimates readily available. Distribution in this way encourages State universities and State agencies to use data that are comparable for all States and counties and that are consistent with national totals; using comparable and consistent data enhances the uniformity of analytic approaches taken in economic development programs and improves the recipients' ability to assess local area economic developments and to service their local clientele.

3.3 BEA Annual State and County Personal Income Estimation Methodology

3.3.1 Overview

The following discussion will focus on the annual estimates of State and county personal income. BEA's quarterly State personal income, annual State disposable personal income, annual State and county full- and part-time employment, and gross State product (GSP) estimates are produced in a manner similar to those described below. (The methodologies for quarterly State personal income and for annual State disposable personal income are described in BEA (1989, pp. M-32-37); the methodology for GSP is described in BEA (1985) and in Trott, Dunbar, and Friedenberg (1991, pp. 43-45).

The personal income of an area is defined as the income received by, or on behalf of, all the residents of the area. It consists of the income received by persons from all sources, that is, from participation in production, from both government and business transfer payments, and from government interest. Personal income is measured as the sum of wage and salary disbursements, other labor income, proprietors' income, rental income of persons, personal dividend income, personal interest income, and transfer payments, less personal contributions for social insurance. Per capita personal income is measured as the personal income of the residents of an area divided by the resident population of the area.

At the national level, personal income is part of the personal income and outlay account, which is one of five accounts in a set that constitutes the national income and product accounts. Such accounts do not now exist below the national level; however, personal income has long been available for States and local areas. In addition, GSP, which corresponds to the national measure gross domestic product, and some elements of personal outlays (personal tax and nontax payments) are available for States but not for local areas. GSP is estimated separately from State personal income, but the two measures share most of the elements of wages and salaries, other labor income, and proprietors' income by State of work. For a tabular representation of the relationships among gross domestic product, State earnings, and GSP, see Table 2 in Trott et. al. (1991, p. 44).

3.3.2 Differences Between the National and Subnational Estimates

The definitions underlying the State and local area estimates of personal income are essentially the same as those underlying the national estimates of personal income. However, the national estimates of personal income include the labor earnings (wages and salaries and other labor income) of residents of the United States temporarily working abroad, whereas the subnational estimates include the labor earnings of persons residing only in the 50 States and the District of Columbia. Specifically, the national estimates include the labor earnings of Federal civilian and military personnel stationed abroad and of residents who are employed by U.S. firms and are on temporary foreign assignment. An "overseas" adjustment is made to exclude the labor earnings of these workers from the national totals before the totals are used as controls for the State estimates.

An important classification difference between national and subnational estimates relates to border workers--that is, residents of the United States who work in adjacent countries (such as Canada) and foreigners who work in the United States but who reside elsewhere. At the national level, the net flow of the labor earnings of border workers and the labor earnings of U.S. residents employed by international organizations and by foreign embassies and consulates in the United States are included in the measurement of the "rest-of-the-world" sector. At the State and local area levels, however, only the labor earnings of U.S. residents employed by international organizations and by foreign embassies and consulates in the United States are treated as a component of personal income. Border workers are treated as commuters, and their earnings flows are reflected in personal income through the residence adjustment procedures.

Statistical differences between the national and subnational series may reflect the different estimating schedules for the two series. The State and local area estimates usually incorporate source data that are not available when the national estimates are prepared. The national estimates are usually revised the following year to reflect the more current State and local area data.

3.3.3 Sources of Data

BEA uses information collected by others to prepare its estimates of State and local area personal income. Generally, two kinds of information are used to measure the income of persons: Information generated at the point of disbursement of the income and information elicited from the recipient of the income. The first kind is data drawn from the records generated by the administration of various Federal and State government programs; the second kind is survey and census data.

The following are among the more important sources of the administrative record data: The State unemployment insurance programs of the Employment and Training Administration, Department of Labor; the social insurance programs of the Social Security Administration and the Health Care Financing Administration, Department of Health and Human Services; the Federal income tax program of the Internal Revenue Service, Department of the Treasury; the veterans benefit programs of the Department of Veterans Affairs; and the military payroll systems of the Department of Defense. The two most important sources of census data are the censuses of agriculture and of population. (BEA uses little survey data to prepare the State and local area estimates; however, the Department of Agriculture makes extensive use of surveys to prepare the State farm income estimates and the county cash receipts and crop production estimates that BEA uses in the derivation of the farm income components of personal income.) The data obtained from administrative records and censuses are used to estimate about 90 percent of personal income. Data of lesser scope and relevance are used for the remaining 10 percent.

When data are not available in time to be incorporated into the current estimating cycle, interim estimates are prepared using the previous year's State or county distribution. The interim estimates are revised during the next estimating cycle to incorporate the newly available data.

Using data that are not primarily designed for income measurement has several advantages and disadvantages. Using administrative record data and census data, BEA can prepare the estimates of State and local area personal income on an annual basis, in considerable detail, at relatively low cost, and without increasing the reporting burden of businesses and households. However, because these data are not designed primarily for income measurement, they often do not precisely "match" the series being estimated and must be adjusted to compensate for differences in content (definition and coverage) and geographic detail.

3.3.4 Controls and the Allocation Procedure

The national estimates for most components of wages and salaries and transfer payments, which together account for about 75 percent of personal income, are based largely on the sum of subnational source data, and the procedure used to prepare the State and county estimates causes only minor changes to the source data. For other components of personal income, either detailed geographic coding is not available for all source data, or more comprehensive and more reliable information is available for the Nation than for States and counties. For these reasons, the estimates of personal income are first constructed at the national level. The subnational estimates are constructed as elements of the national totals, using the subnational data. Thus, the national estimates, with some adjustment for definition, serve as the "control" for the State estimates, and the State estimates, in turn, serve as controls for the county estimates.

The State estimates are made by allocating the national total for each component of personal income to the States in proportion to each State's share of a related economic series. Similarly, the county estimates are made, in somewhat less component detail, by allocating the State total. In some cases, the related series used for the allocation may be a composite of several items (e.g., wages, tips, and pay-in-kind) or the product of two items (e.g., average wages times the number of employees). In every case, the final estimating step for each income estimate is its adjustment to the appropriate higher level total. This procedure is called the allocation procedure.

The allocation procedure, as used to estimate a component of State personal income, is given by

$$Y_s = (Y_n)\left(\frac{X_s}{X_n}\right)$$

where $Y_s =$ estimator of the income component for State s

          $Y_n =$ estimate of the income component for the Nation (used as a control total for the State estimates)

          $X_s =$ source datum related to the income component for State s

$X_n =$ national sum of State source data relating to the income component
$(X_n = \Sigma X_s).$

For the cases in which the national estimate is derived as the sum of the State data plus an additional amount A_n for which State data are unavailable, the allocation procedure can be restated in two steps as

$$A_s = (A_n)\left(\frac{X_s}{X_n}\right)$$

$$Y_s = X_s + A_s$$

where A_s is the State estimator of that portion of Y for which State data are not available. In effect, Y_s is a composite estimator consisting of X_s, the best possible direct estimator (100 percent sample) of that portion of Y for which State data are available, plus A_s, the domain-indirect estimator of the remainder of Y.

The source data that underlie the national estimates are frequently more timely, detailed, and complete than the available State and county data. The use of the allocation procedure imparts some of these aspects of the national estimates to the subnational estimates and allows the use of subnational data that are related but that do not always precisely match the series being estimated. The use of this procedure also yields an additive system wherein the county estimates sum to the State totals and the State estimates sum to the national total.

3.3.5. Place of Measurement

Personal income, by definition, is a measure of income received; therefore, estimates of State and local area personal income should reflect the residence of the income recipients. However, the data available for regional economic measurement are frequently recorded by the recipients' place of work. The data underlying the estimates can be viewed as falling in four groups according to the place of measurement.

(1) For the estimates of wages and salaries, other labor income, and personal contributions for social insurance by employees, most of the source data are reported by industry in the State and county in which the employing establishment is located; therefore, these data are recorded by place of work. The estimates based on these data are subsequently adjusted to a place-of-residence basis for inclusion in the personal income measure.

(2) For nonfarm proprietors' income and personal contributions for social insurance by the self-employed, the source data are reported by tax-filing address. These data are largely recorded by place of residence.

(3) For farm proprietors' income, the source data are reported and recorded at the principal place of production, which is usually the county in which the farm has most of its land.

(4) For military reserve pay, rental income of persons, personal dividend income, personal interest income, transfer payments, and personal contributions for supplementary medical insurance and for veterans life insurance, the source data are reported and recorded by the place of residence of the income recipients.

3.3.6 Sources and Methods for Annual State and County Income Estimates

3.3.6.1 Framework

Personal income is estimated as the sum of its detailed components; the major types of payments that comprise those components are shown in Table 3.2, together with the related percents of personal income and the principal sources of data used to estimate the components. The following methodology presentation consists of a section for each of the six types of payment and a section for the residence adjustment. The methodologies for some types of payment and for many of the individual income components are omitted from this presentation, but a complete presentation is available (BEA 1991, pp. M-7-27).

3.3.6.2 Wage and salary disbursements

Wage and salary disbursements, which accounted for about 58 percent of total personal income at the national level in 1990, are defined as the monetary remuneration of employees, including the compensation of corporate officers; commissions, tips, and bonuses; and receipts in kind that represent income to the recipient. They are measured before deductions, such as social security contributions and union dues. The estimates reflect the amount of wages and salaries disbursed during the current period, regardless of when they were earned.

The following description of the procedures used in making the estimates of wage and salary disbursements is divided into three sections: Wages and salaries that are covered under the unemployment insurance (UI) program, wages and salaries that are not covered under the UI program, and wages and salaries that are paid in kind.

Table 3.2 Sources of the Estimation for Local Area Personal Income

Component of Personal Income	Percent of Personal Income [a]	Sources of subnational data used to allocate national control totals (see section 3.3.4)
Wages and Salaries:		
Private	47.55	Annual ES-202 wages and salaries.
Federal Civilian............	2.16	Annual ES-202 wages and salaries.
Federal Military............	0.98	DOD payroll outlays.
State and Local Government................	7.62	Annual ES-202 wages and salaries.
Other Labor Income........	5.53	BEA wages and salaries and employment; other agency sources.
Proprietors' Income:		
Nonfarm.....................	7.56	IRS, CBP and AMA.
Farm...........................	1.08	USDA and Census Bureau.
Dividends, Interest, and Rental Income Received by Persons..................	17.39	IRS.
Transfer Payments...........	14.98	SSA, HCFA, Census Bureau, DVA, and other agency sources.
Personal Contributions for Social Insurance.......	4.83	SSA and Census Bureau.
Residence Adjustment........	-- [b]	IRS and Census Bureau.

a. All percentages are shares of U.S. total personal income for 1990. Because personal contributions for social insurance are deducted in the derivation of personal income, the sum of the percentages exceeds 100 percent by about twice the personal contributions percentage.

b. At the national level, the residence adjustment is negligible; at the county level, the absolute value of the residence adjustment constitutes, on average, about 12.5 percent of total personal income.

Abbreviations: AMA (American Medical Association), CBP (County Business Patterns, published by the Census Bureau), DOD (Department of Defense), DVA (Department of Veterans' Affairs), ES-202 (Tabulations of wages reported on employers' unemployment insurance tax returns, provided to BEA by the Bureau of Labor Statistics), IRS (Internal Revenue Service), HCFA (Health Care Financing Administration), USDA (U.S. Department of Agriculture), SSA (Social Security Administration).

Wages and salaries covered by the UI program

The estimates of about 95 percent of wages and salaries are derived from tabulations by the State employment security agencies (ESA's) from their State employment security reports (form ES-202). These tabulations summarize the data from the quarterly UI contribution reports filed with a State ESA by the employers subject to that State's UI laws. Employers usually submit reports for each "county reporting unit"--i.e., for the sum of all the employer's establishments in a county for each industry. However, in some cases, an employer may group very small establishments in a single "statewide" report without a county designation. Each quarter, the various State ESA's submit the ES-202 tabulations on magnetic tape to the Bureau of Labor Statistics (BLS), which provides a duplicate tape to BEA. The tabulations present monthly employment and quarterly wages for each county in Standard Industrial Classification four-digit detail. (The ES-202 tabulations through 1987 reflect the 1972 SIC, and those for 1988-92, the 1987 SIC.) Under the reporting requirements of most State UI laws, wages include bonuses, tips, gratuities, and the cash value of meals and lodging supplied by the employer.

The BEA estimates of wage and salary disbursements are made, with a few exceptions, at the SIC two-digit level. However, the availability of the ES-202 data in SIC four-digit detail facilitates the detection of errors and anomalies; this detail also makes it possible to isolate those SIC three-digit industries for which UI coverage is too incomplete to form a reliable basis for the estimates. In this case, the SIC two-digit estimate is prepared as the sum of two pieces: The fully covered portion, which is based on the ES-202 data, and the incompletely covered portion, which is estimated as described in the section on wages and salaries not covered by the UI program.

The ES-202 wage and salary data do not precisely meet the statistical and conceptual requirements for BEA's personal income estimates. Consequently, the data must be adjusted to meet the requirements more closely. The adjustments affect both the industrial and geographic patterns of the State and county UI-based wage estimates.

Adjustment for statewide reporting.--Wages and salaries reported for statewide units are allocated to counties in proportion to the distribution of the wages and salaries reported by county; the allocations for each State are made for each private-sector industry (generally at the SIC two-digit level) and for five government components.

Adjustment for industry nonclassification.--The industry detail of the ES-202 tabulations regularly shows minor amounts of payroll that have not been assigned to any industry. For each State and county, the amount of ES-202 payrolls in this category is distributed among the industries in direct proportion to the industry-classified payrolls.

Misreporting adjustment.--This adjustment--the addition of estimates of wages and salaries subject to UI reporting that employers do not report--is made to the ES-202 data for all covered private-sector industries. At the national level, the estimate for each industry is made in two parts--one for the underreporting of payrolls on UI reports filed by employers and one for the payrolls of employers that fail to file UI reports (Parker, 1984). The source data necessary to replicate this methodology below the national level are not available. Instead, the national adjustment for each industry is allocated to States and counties in proportion to ES-202 payrolls.

Adjustments to government components.--Alternative source data are substituted for the ES-202 data when the latter series reflects excessively large proportions of Federal civilian payrolls that are not reported by county or of State government payrolls that are apparently reported in the wrong counties. For Federal civilian wages and salaries, the alternative source data are tabulations of employment by agency and county prepared by the Office of Personnel Management. For State government wages and salaries, the alternative source data are place-of-work wage data derived from an unpublished tabulation of journey-to-work (JTW) data from the decennial Census of Population. (All income estimates using decennial Census of Population data were updated to incorporate 1990 Census of Population data in a regional comprehensive revision released in the spring of 1994.)

Adjustments for noncovered elements of UI-covered industries.--BEA presently makes adjustments for the following noncovered elements:
   o   Tips;
   o   Commissions received by insurance solicitors and real estate agents;
   o   Payrolls of electric railroads, railroad carrier affiliates, and railway labor organizations;
   o   Salaries of corporate officers in Washington State;
   o   Payrolls of nonprofit organizations exempt from UI coverage because they have fewer than four employees;
   o   Wages and salaries of students employed by the institutions of higher education in which they are enrolled;
   o   Allowances paid to Federal civilian employees in selected occupations for uniforms; and
   o   Salaries of State and local government elected officials and members of the judiciary.

Except for tips, these elements are exempted from State UI coverage. Tips are covered by the various UI laws. BEA assumes that this form of income payment is considerably underreported, and it therefore makes additional estimates of tips in industries where tipping is most customary.

National and State estimates of each of the noncovered elements are made (based on either direct data or indirect indicators). These estimates are added to the ES-202 payroll amount for the industry of the noncovered element to produce the final estimates for that industry. Because of the lack of relevant data, county estimates are made by allocating the final State total by the distribution of ES-202 payrolls for the appropriate industry.

Wages and salaries not covered by the UI program

Eight industries are treated as noncovered in making the State and county estimates of wage and salary disbursements: (1) Farms, (2) farm labor contractors, (3) railroads, (4) private elementary and secondary schools, (5) religious membership organizations, (6) private households, (7) military, and (8) "other." The estimates for these industries are based on a variety of sources. For example, the estimates for railroads ar based mainly on employment data provided by the Association of American Railroads, and the estimates for the military services are based mainly on payroll data provided by the Department of Defense.   See BEA (1991) for the methodology for the noncovered industries.

Wages and salaries paid in kind

The value of food, lodging, clothing, and miscellaneous goods and services furnished to employees by their employers as payment, in part or in full, for services performed is included in the wage and salary component of personal income and is referred to as "pay-in-kind." The estimates for UI-covered industries are prepared as an integral part of total wages and salaries for those industries, based on the ES-202 data. The estimates for most on the noncovered industries are based on pertinent employment data. See BEA (1991) for the methodology for pay-in-kind.

3.3.6.3 Other labor income

Other labor income (OLI), which accounted for about 5.5 percent of total personal income at the national level in 1990, consists primarily of employer contributions to private pension and welfare funds; these employer contributions account for approximately 98 percent of OLI. The "all other" component of OLI consists of directors' fees, judicial fees, and compensation of prisoners. Employer contributions for social insurance, which are paid into government-administered funds, are not included in OLI; under national income and product account conventions, it is the benefits paid from social insurance funds--which are classified as transfer payments--that are measured as part of personal income, not the employer contributions to the funds.

Employer contributions to private pension and welfare funds

Private pension and profit-sharing funds, group health and life insurance, and supplemental unemployment insurance.--The larger part of the national estimates of employer contributions to private pension and welfare funds is developed from Internal Revenue Service tabulations of data from proprietorship and corporate income tax returns published in Statistics of Income. However, these data are not suitable for making the subnational estimates because most multiestablishment corporations file tax returns on a companywide basis instead of for each establishment and because the State in which a corporation's principal office is located is often different from the State of its other establishments. As a result, the geographic distribution of the data tabulated from the tax returns does not necessarily reflect the place of work of the employees on whose behalf the contributions are made.

For private-sector employees, the State and county estimates of employer contributions to private pension and profit-sharing funds, group health and life insurance, and supplemental unemployment insurance are made, for all types of employer contributions combined, at the SIC two-digit level, the same level of industrial detail as the wage and salary estimates. The national total of employer contributions for each industry is allocated to the States and counties in proportion to the estimates of wage and salary disbursements for the corresponding industry. The use of subnational wage estimates to allocate the national estimates of employer contributions to private pension and welfare funds is based on the assumption that the relationship of contributions to payrolls for each industry is the same at the national, State, and county levels. The procedure reflects the wide variation in contribution rates--relative to payrolls--among industries (and therefore reflects appropriately the various mixes of industries among States and counties). It does not reflect the variation in contribution rates among States and counties for a given industry.

The Federal Government makes contributions to a private pension fund, called the Thrift Savings Plan, on behalf of its civilian employees who participate in the Federal Employees Retirement System (mainly employees hired after 1983). In the absence of direct data below the national level, the national estimate is allocated to States and counties in proportion to the estimates of Federal civilian wages and salaries.

State government contributions to private pension plans consist of annuity payments made by State governments on behalf of selected employee groups--primarily teachers. The State estimates are based on direct data from the Teachers Insurance and Annuity Association/College Retirement Equities Fund. The county estimates are prepared by allocating the State estimates in proportion to the estimates of State and local government education wages and salaries.

In the absence of direct data below the national level, the national estimates of Federal, State, and local government contributions to private welfare funds on behalf of their employees are allocated to States and counties in proportion to ES-202 employment data for each level of government.

Privately administered workers' compensation.--The State estimates for this subcomponent are based mainly on direct data provided by the National Council on Compensation Insurance and by the Social Security Administration; the county estimates for each SIC two-digit industry reflect the geographic distribution of wages and salaries. The methodology for this income component is given in BEA (1991).

"All other" OLI

The methodology for "all other" OLI--primarily directors' fees and jury and witness fees--is given in BEA (1991). The State and county estimates for directors' fees--the largest of these subcomponents--reflect the geographic distribution of wages and salaries in each industry.

3.3.6.4 Proprietors' Income

Proprietors' income, which accounted for about 8.5 percent of total personal income at the national level in 1990, is the income, including income-in-kind, of sole proprietorships and partnerships and of tax-exempt cooperatives. The imputed net rental income of owner-occupants of farm dwellings is included. Dividends and monetary interest received by proprietors of nonfinancial business, monetary rental income received by persons who are not primarily engaged in the real estate business, and the imputed net rental income of owner-occupants of nonfarm dwellings are excluded; these incomes are included in dividends, net interest, and rental income of persons. Proprietors' income, which is treated in its entirety as received by individuals, is estimated in two parts--nonfarm and farm.

Nonfarm proprietors' income

Nonfarm proprietors' income is the income received by nonfarm sole proprietorships and partnerships and by tax-exempt cooperatives. The State and county estimates of the income of sole proprietors and partnerships for all but three of the SIC two-digit industries are based on 1981-83 tabulations from Internal Revenue Service (IRS) form 1040, Schedule C (for sole proprietors), and form 1065 (for partnerships). Tabulations either of gross receipts or of profit less loss from the two forms combined are used either to attribute a national total to the States or as direct data. Two national totals are used for each industry: One for income reported on the income tax returns--as adjusted to conform with national income and product accounting conventions--and one for an estimate of the income not reported on tax returns.

For the adjustments for unreported income, no direct data are available below the national level. The national total for each industry is attributed to States in proportion to the IRS State distribution of gross receipts for the industry. For the reported portion of nonfarm proprietors' income, the State estimates for each of 45 industries are based on the IRS distribution of profit less loss for the industry, and the estimates for each of another 20 industries (together accounting for 3 percent of total nonfarm proprietors' income) are based on the IRS distribution of gross receipts for the industry. For the latter group, the IRS distribution of profit less loss, although preferable in concept, is not used as a basis for State estimates because the extreme year-to-year volatility of the State data suggests that they are unreliable.

The 1983 State estimates prepared by the foregoing methodology are extended to later years based mainly on the number of small establishments in each industry as determined from the Census Bureau's County Business Patterns; see BEA (1989) for a full description of the methodology.

For the three remaining industries, limited partners' income presents a special estimating problem. In these industries--crude petroleum and natural gas extraction, real estate, and holding and investment companies--limited partnerships are often used as tax shelters. Limited partners' participation in partnerships is often purely financial; their participation more closely resembles that of investors than that of working partners. Accordingly, the usual assumption that the State from which the partnership files its tax return is the same as the residence of the individual partners is unsatisfactory. No direct data on the income of partners by their place of residence are available. The national estimates of proprietors' income for these industries are attributed to States in the same proportion as dividends received by individuals (based on all-industry dividends reported on IRS form 1040).

The State estimates of the income of tax-exempt cooperatives are based on data provided by the Rural Electrification Administration (for electric and telephone cooperatives) and the Agricultural Cooperative Service (for farm supply and marketing cooperatives); see BEA (1989) for the methodology.

The methodology for the county estimates of nonfarm proprietors' income is similar to the State methodology, but less direct data are used for many industries because problems with data volatility are greater at the county level. See BEA (1991) for a full description of the county methodology.

Farm proprietors' income

The estimation of farm proprietors' income starts with the computation of the realized net income of all farms, which is derived as farm gross receipts less production expenses. This measure is then modified to reflect current production through a change-in-inventory adjustment and to exclude the income of corporate farms and salaries paid to corporate officers. Tables showing the derivation of State and county farm proprietors' income in detail are available from the Regional Economic Information System.

The concepts underlying the national and State BEA estimates of farm income are generally the same as those underlying the national and State farm income estimates of the U.S. Department of Agriculture (USDA). The major definitional difference between the two sets of estimates relates to corporate farms. The USDA totals include net income of corporate farms, whereas the BEA personal income series, which measures farm proprietors' net income, by definition excludes corporate farms. Additionally, BEA classifies the salaries of officers of corporate farms as part of farm wages and salaries; USDA treats the corporate officers' salaries as returns to corporate ownership and as part of the total return to farm operators.

The State control totals for the BEA county estimates of farm proprietors' income are taken from the component detail of the USDA State estimates, which are modified to reflect BEA definitions and to include interfarm intrastate sales.

The methods used to estimate farm proprietors' income at the county level rely heavily on data obtained from the 1974, 1978, 1982, and 1987 censuses of agriculture and on selected annual county data prepared by the State offices affiliated with the National Agricultural Statistics Service (NASS), USDA. The NASS data, which are described by Iwig in Chapter 7 of this report, are used, wherever possible, to interpolate and extrapolate to noncensus years. In addition, data from other sources within USDA, such as the Agricultural Stabilization and Conservation Service, are used to prepare a fairly detailed income and expense statement covering all farms in the State and county.

For census years, BEA prepares county estimates of 46 components of gross income and 13 categories of production expenses. For intercensal and postcensus years, the component detail of the estimates for each State is set to take advantage of the best annual county data available for the State.

Farm gross income includes estimates for the following items: (1) The cash receipts from farm marketing of crops and livestock (in component detail); (2) the income from other farm-related activities, including recreational services, forest products, and custom-feeding services performed by farm operators; (3) the payments to farmers under several government payment programs; (4) the value of farm products produced and consumed on farms; (5) the gross rental value of farm dwellings; and (6) the value of the net change in the physical volume of farm inventories of crops and livestock.

Cash receipts from marketing is the most important component of farm gross income. The USDA generally has annual production, marketing, and price data available for preparing the State estimates for about 150 different commodities. However, annual county estimates of cash receipts--usually for total crops and for total livestock--are currently available for only 19 States (BEA 1991, fn. 15, p. M-14). For the other States, the USDA State estimates of cash receipts from the marketing of individual commodities are summed into the 13 crop and 5 livestock groups for which value-of-sales data are reported by county in the censuses of agriculture. The aggregates for the census years are then allocated by the related census county distributions. Estimates for intercensal years are based on supplemental county estimates of annual production of selected field crops and on State season average prices available from the State NASS offices, or they are calculated by straight-line interpolation between the census years and adjusted to State USDA levels.

The county estimates of the remaining components of gross income, of production expenses, of the adjustment for interfarm intrastate transactions, and of the adjustment to exclude the income of corporate farms are based mainly on data from the censuses of agriculture and data provided by NASS and by the Agricultural Stabilization and Conservation Service. See BEA (1991) for a full description of the methodology.

3.3.6.5 Personal Dividend Income, Personal Interest Income, and Rental Income of Persons

These components accounted for more than 17 percent of total personal income in 1990. Dividends are payments in cash or other assets, excluding stock, by corporations organized for profit to noncorporate stockholders who are U.S. residents. Interest is the monetary and imputed interest income of persons from all sources. Imputed interest represents the excess of income received by financial intermediaries from funds entrusted to them by persons over income disbursed by these intermediaries to persons. Part of imputed interest reflects the value of financial services rendered without charge to persons by depository institutions. The remainder is the property income held by life insurance companies and private noninsured pension funds on the account of persons; one example is the additions to policyholder reserves held by life insurance companies.

Rental income of persons consists of the monetary income of persons (except those primarily engaged in the real estate business) from the rental of real property (including mobile homes); the royalties received by persons from patents, copyrights, and rights to natural resources; and the imputed net rental income of owner-occupants of nonfarm dwellings.

The State and county estimates of dividends, interest, and rent are based mainly on data tabulated from Federal individual income tax returns by the Internal Revenue Service. The methodology for dividends, interest, and rent is given in BEA (1991).

3.3.6.6 Transfer payments

Transfer payments are payments to persons, generally in monetary form, for which they do not render current services. As a component of personal income, they are payments by government and business to individuals and nonprofit institutions. Nationally, transfer payments accounted for almost 15 percent of total personal income in 1990. At the county level, approximately 75 percent of total transfer payments are estimated on the basis of directly reported data. The remaining 25 percent are estimated on the basis of indirect, but generally reliable, data.

For the State and county estimates, approximately 50 subcomponents of transfer payments are independently estimated using the best data available for each subcomponent. The methodology for all of these subcomponents is given in BEA (1991); the following items are presented here as examples.

Old-age, survivors, and disability insurance (OASDI) payments.--These payments, popularly known as social security, consist of the total cash benefits paid during the year, including monthly benefits paid to retired workers, dependents, and survivors and special payments to persons 72 years of age and over; lump-sum payments to survivors; and disability payments to workers and their dependents. The State estimates of each OASDI segment are based on Social Security Administration (SSA) tabulations of calendar year payments. The county estimates of total OASDI benefits are based on SSA tabulations of the amount of monthly benefits paid to those in current-payment status on December 31, by county of residence of the beneficiaries.

Medical vendor payments.--These are mainly payments made through intermediaries for care provided to individuals under the federally assisted State-administered medicaid program. Payments made under the general assistance medical programs of State and local governments are also included. The county estimates are based on available payments data from the various State departments of social services. For States where no county data are available, the county estimates are based on the distribution of payments made under the aid to families with dependent children program.

Aid to families with dependent children (AFDC).--This State-administered program receives Federal matching funds to provide payments to needy families. The State estimates are based on unpublished quarterly payments data provided by the SSA. The county estimates are prepared from payments data provided by the various State departments of social services. County data are no longer being received from some States; for these States, the most recent available data are used for the county estimates for each subsequent year.

State unemployment compensation.--These are the cash benefits, including special benefits authorized by Federal legislation for periods of high unemployment, from State-administered unemployment insurance (UI) programs. Most States report benefits directly by county, but a few report by local district office. In the latter case, local district office data are distributed among the counties within the jurisdiction of the local district office in proportion to the annual average number of unemployed persons estimated by the Bureau of Labor Statistics (BLS). When the State is unable to supply the county data in time to meet the publication deadline, a preliminary set of estimates is made and is revised the following year to incorporate the delayed county data. The preliminary county estimates are prepared by extrapolating the preceding year's estimates forward by the change in the BLS estimate of the annual average number of unemployed persons.

Veterans life insurance benefit payments.--These are the claims paid to beneficiaries and the dividends paid to policyholders from the five veterans life insurance programs administered by the Department of Veterans Affairs. The county allocations of the combined payments of death benefits and dividends are based on the distribution of the veteran population.

Interest payments on guaranteed student loans.--These are the payments to commercial lending institutions on behalf of individuals who receive low-interest deferred-payment loans from these institutions to pay the expenses of higher education. The State estimates are based on Department of Education data on the number of persons enrolled in institutions of higher education. The county allocations are based on the distribution of the civilian population.

3.3.6.7 Personal Contributions For Social Insurance

Personal contributions for social insurance are the contributions made by individuals under the various social insurance programs. These contributions are excluded from personal income by treating them as explicit deductions. Payments by employees and the self-employed for social security, medicare, and government employees' retirement are included in this component. Also included are the contributions that are made by persons participating in the veterans life insurance program and in the supplementary medical insurance portion of the medicare program. The State and county estimates of personal contributions for social insurance are generally based either on direct data from the administering agency or on the geographic distribution of the appropriate earnings component; see BEA (1989 and 1991) for the full methodologies.

3.3.6.8 Residence Adjustment

Personal income is a "place-of-residence" measure of income, but the source data for the components that compose more than 60 percent of personal income are recorded by place of work. The adjustment of the estimates of these components to a place-of-residence basis is the subject of this section.

At the national level, place of residence is an issue only for border workers (mainly those living in the United States and working in Canada or Mexico and vice versa). At the State and county levels, the issue of place of residence is more significant. Individuals commuting to work between States are a major factor where metropolitan areas extend across State boundaries--for example, the Washington, DC-MD-VA MSA. Individuals commuting between counties are a major factor in every multicounty metropolitan area and in many nonmetropolitan areas.

BEA's concept of residence as it relates to personal income refers to where the income to be measured is received rather than to "usual," "permanent," or "legal" residence. It differs from the Census Bureau's concept mainly in the treatment of migrant workers. The decennial census counts many of these workers at their usual place of residence rather than where they are on April 1 when the census is taken. Except for out-of-State workers in Alaska (where migrant workers are unusually important) and for certain groups of border workers, BEA assigns the wages of migrant workers to the area in which they reside while performing the work. Similarly, BEA assigns the income of military personnel to the county in which they reside while on military assignment, not to the county in which they consider themselves to be permanent or legal residents. Thus, in the State and local area personal income series, the income of military personnel on foreign assignment is excluded because their residence is outside of the territorial limits of the United States.

Three of the six major components of personal income are recorded, or are treated as if recorded, on a place-of-residence (where-received) basis. They are transfer payments; personal dividend income, personal interest income, and rental income of persons; and proprietors' income. Nonfarm proprietors' income is treated as income recorded on a place-of-residence basis because the source data for almost all of this part of proprietors' income are reported to the IRS by tax-filing address, which is usually the filer's place of residence. The source data for farm proprietors' income are recorded by place of production, which is usually in the same county as the proprietor's place of residence.

The remaining three major components--wages and salaries, other labor income (OLI), and personal contributions for social insurance--are estimated, with minor exceptions, from data that are recorded by place of work (point of disbursement). The sum of these components (wages plus OLI minus contributions) is referred to as "income subject to adjustment" (ISA).

Residence adjustment procedure (excluding border workers)

The county residence adjustment estimates for 1981 and later years are based on those for 1980 and 1990 because intercounty commuting data are available only from the decennial censuses of population. (Data from the 1990 Census of Population were introduced into the residence adjustment estimates as part of the comprehensive revisions to the State and local area personal income estimates completed in the Spring of 1994.) The estimation of these adjustments can be understood using the example of a two-county area comprising counties f and g. The two-county example is easily generalized to more complex situations.

For 1990, the initial residence adjustment estimate for county f (RA_f) was calculated as total 1990 inflows of earnings to f from g (IN_f) minus total 1990 outflows of earnings from f to g (OUT_f):

$$RA_f = IN_f - OUT_f.$$

The estimates of IN_f and OUT_f were prepared from industry-level data. The share $(I_{f,k})$ of total wages or OLI in a particular industry k in g that was earned by residents of f was used in the estimation of industry-level inflows to f. Analogously, the share $(O_{f,k})$ of wages or OLI in a particular industry in f that was earned by residents of g was used in the estimation of industry-level outflows from f. Both $I_{f,k}$ and $O_{f,k}$ were calculated from journey-to-work (JTW) data from the 1990 Census of Population on the number of wage and salary workers (W) and on their average earnings (E) by county of work for each county of residence:

$$I_{f,k} = \frac{income\ earned\ in\ g\ by\ residents\ of\ f}{total\ income\ earned\ in\ g}$$

$$= \frac{(W_{(f \to g),k})(E_{(f \to g),k})}{(W_{(f \to g),k})(E_{(f \to g),k}) + (W_{(g \to g),k})(E_{(g \to g),k})}$$

$$O_{f,k} = \frac{\text{income earned in } f \text{ by residents of } g}{\text{total income earned in } f}$$

$$= \frac{(W_{(g\to f),k})(E_{(g\to f),k})}{(W_{(g\to f),k})(E_{(g\to f),k}) + (W_{(f\to f),k})(E_{(f\to f),k})}.$$

(Where two subscripts are used with an arrow, the first identifies place of residence, and the second identifies place of work. For example, $W_{(f\to g),k}$ is the number of workers in industry k who lived in f but worked in g.)

Industry-level inflows to county f from county g ($IN_{f,k}$) were derived by multiplying the inflow ratio by the income subject to adjustment in industry k in g ($ISA_{g,k}$); industry-level outflows from f to g ($OUT_{f,k}$) were derived by multiplying the outflow ratio by the income subject to adjustment in industry k in f ($ISA_{f,k}$):

$$IN_{f,k} = (I_{f,k})(ISA_{g,k})$$
$$OUT_{f,k} = (O_{f,k})(ISA_{f,k}).$$

(For the private sector, adjustment factors and gross flows were calculated at the SIC division level; see also BEA (1991, fn. 22, p. M-22).) Summing over all industries yields total inflows to f and total outflows from f:

$$IN_{f.} = \sum_{k=1}^{N} IN_{f,k}$$

$$OUT_{f.} = \sum_{k=1}^{N} OUT_{f,k}$$

The initial 1990 BEA estimates were modified in three situations. First, for clusters of counties identified as being closely related by commuting (mostly multicounty metropolitan areas), modifications were made to incorporate the 1989 wage and salary distribution from the 1990 Census of Population. (The 1989 wage and salary distribution from the 1990 Census of Population reflects the residential distribution of the income recipients as of April 1, 1990, regardless of where they were living when they received the wages and salaries.) These modifications are needed because in numerous cases the 1990-census JTW data and the source data for the BEA wage estimates are inconsistently coded by place of work. (For example, the source data may attribute too much of the wages of a multiestablishment firm to the county of the firm's main office, or the geographic coding of the Defense Department payroll data and of the JTW data may attribute a military base extending across county boundaries to different counties.) Initial county estimates of place-of-residence wages and salaries were derived as place-of-work wages and salaries plus net residence adjustment for wages and salaries. (For the calculation of this net residence adjustment, only the gross flows for wages and salaries were used.) Then, the initial 1990 BEA place-of-residence wage and salary estimates were summed to a total for each cluster. Finally, the BEA total for each cluster was redistributed among the counties of the cluster in the same proportion as the 1989 wage and salary distribution from the 1990 census. To facilitate the extension of the 1990 residence adjustment estimates to later years, the cluster-based modifications--derived as net additions to or subtractions from the initial residence adjustment estimates for each of the 1,345 counties--were expressed as gross flows between pairs of counties within the same cluster. In the simplest case--a two-county cluster--the additional gross flow was assumed to be from the county with the negative modification to the county with the (exactly offsetting) positive modification.

Second, modifications were made for selected noncluster adjacent counties if large, offsetting differences occurred between the initial 1990 BEA estimates and the census wage data for these counties. These adjacent-county modifications were expressed as gross flows in the same way and for the same reason as the cluster-based modifications.

Third, modifications were made for eight Alaska county equivalents (boroughs and census areas) to reflect the large amounts of labor earnings received by seasonal workers from out of State. The 1990-census JTW data reflect the "commuting" of many of these workers, and the initial 1990 residence adjustment estimates for a majority of the county equivalents did not require modifications. However, for eight county equivalents, the initial 1989 estimates yielded BEA place-of-residence wage and salary totals that were so much higher than the comparable census data that they could not be an accurate reflection of the wages of only the permanent residents. (The 1989 residence adjustment estimates, although based mainly on the 1990-census JTW data, also reflect--at the appropriate one-tenth weight--1980-census JTW data.) Based on the assumption that the excess amounts were attributable to out-of-State migrant workers, these amounts were removed by judgmentally increasing the JTW-based gross flows to the large metropolitan counties of Washington, Oregon, and California.

For 1991 and later years, the 1990 estimates of total inflows (IN_f^{1990}; denoted previously without the superscript) and outflows by industry ($OUT_{f,k}^{1990}$; denoted previously without the superscript) are extrapolated. (The gross flows that are extrapolated consist of those calculated with the 1990-census JTW data, those devised to express the cluster-based and adjacent-county modifications, and those derived from the Alaska out-of-State-workers modifications.) Changes in intercounty commuting patterns are incorporated into the estimates by a ratio (CHR_f) in which the numerator is derived from BEA's place-of-work measure of earnings (ISA_f) for all industries and the denominator is derived from an independent place-of-residence measure--wages and salaries reported by individuals to the IRS (IRS_f) (BEA 1991, fn. 27. p. M-22). The ratio for county f in year t (CHR_f^t) is:

$$CHR_f^t = \frac{ISA_f^t / ISA_f^{1990}}{IRS_f^t / IRS_f^{1990}}.$$

Total 1990 inflows to f are extrapolated to year t on the basis of the inverse of CHR_f^t and of the change in IRS_f since 1990:

$$IN_f^t = \left(IN_f^{1990}\right) \left(\frac{IRS_f^t}{IRS_f^{1990}}\right) \left(\frac{1}{CHR_f^t}\right).$$

For each industry, 1990 outflows from f to g are extrapolated to year t on the basis of CHR_f^t and of the change in $ISA_{f,k}$ for the industry since 1990:

$$OUT_{f,k}^t = \left(OUT_{f,k}^{1990}\right) \left(\frac{ISA_{f,k}^t}{ISA_{f,k}^{1990}}\right) \left(CHR_f^t\right).$$

The estimate of net residence adjustment for year t (the final estimate for the noncluster counties and the provisional estimate for the cluster counties) equals total inflows minus total outflows summed over all industries:

$$RA_f^t = IN_{f.}^t - \sum_{k=1}^{N} OUT_{f,k}^t$$

As a last step, the total place-of-residence ISA (ISA plus net residence adjustment) for each cluster is derived and then distributed to the counties of the cluster based on 1990 place-of-residence ISA extrapolated to later years by the percentage change in the IRS-based wage series. The net residence adjustment estimate for each cluster county is calculated as place-of-residence ISA minus place-of-work ISA.

3.4 Evaluation Practices

In the past few years, two major studies were undertaken by BEA to evaluate the State and local area income estimates: (1) a reliability study of the State quarterly personal income series and (2) a study of the accuracy of the county residence adjustment estimates. In addition, in March of this year, the U.S. General Accounting Office (GAO) completed a study of BEA's national and State estimates.

3.4.1 Evaluation of the State Quarterly Personal Income Series

This study provided a detailed measurement and analysis of the reliability of quarterly and annual estimates of State personal income (Brown and Stehle 1990). The study, which covered the State estimates from 1980-87, assessed the reliability of State quarterly personal income using several statistical measures to examine the size of the revisions made to the estimates. One measure used analyzes the range of revisions, where revision is defined as the percent change in the final estimates minus the percent change in the preliminary estimates. Other sets of measures used were dispersion, relative dispersion, bias, and relative bias. The findings of the study were intended to: (1) help BEA isolate particular problem areas in the production of these estimates; and (2) help users of these data determine the suitability for their purposes of the estimates released at different stages of the estimating process. The four principle findings of the study were: (1) the major sources of the revisions to the quarterly percent changes in the preliminary quarterly estimates of State personal income are farm proprietors' income and wages and salaries; (2) largely reflecting wages and salaries, the preliminary quarterly estimates of total personal income tend to be underestimated in fast-growing States and overestimated in slow-growing States; (3) beginning in 1984, the reliability of the second quarterly estimates (that is, the estimates yielded by the first routine revision) was improved by the incorporation of quarterly data from employers' payroll tax reports (the ES-202 data), and (4) the annual revisions of total personal income are smaller than the quarterly revisions.

3.4.2 Residence Adjustment Reliability Study

In October 1988, a study was completed which measured the reliability of the Census commuting data used to prepare the net residence adjustments for county personal income (Zabronsky 1988). While the impact of the residence adjustments are generally small at the region and State level, the residence adjustments constitute a large portion of total personal income for most counties in the U.S. In 1989 for instance, the absolute value of the net residence adjustments accounted for about 12.5 percent of total personal income for all counties, on average, while accounting for about 25 percent of total personal income in metropolitan area counties, on average.

In this residence adjustment reliability study, a comparison of a Census file of county commuting data constructed from the journey-to-work question on the 1980 decennial census was made with a file of aggregate wages and salaries independently tabulated by Census. In the course of the study, comparisons between the journey-to-work and aggregate wage series were explored across a variety of geographic, demographic, and industrial detail to develop a comprehensive reliability profile for the census commuting data.

The major conclusion of this study was that taking the 1980 Census aggregate income series as a benchmark measure of county wages and salary income, the 1980 census journey-to-work data proved to be a highly reliable source for measuring commuter's income in the development of BEA's county residence adjustment estimates. Although careful analysis of the Census journey-to-work wage data did reveal a bias in that series that was correlated with county size, wage imputations undertaken by BEA largely corrected the problem while commuting patterns between counties indicated that for the relevant comparisons, the Census journey-to-work data were consistent with the Census aggregate income-based wage series.

3.4.3 GAO Study of BEA's State and National Estimates

The GAO study (GAO, 1993) was conducted in response to a request by the Honorable Ernest F. Hollings, Chairman, Committee on Commerce, Science, and Transportation, U.S. Senate. Senator Hollings expressed concern about press reports that alleged that BEA "did not incorporate, for political purposes, a downward revision of original employment levels into its October 1991 estimate of first quarter 1991 State personal income growth and its December 1991 estimate of first quarter 1991 gross domestic product (GDP) growth." The report concluded that "We found no evidence that BEA manipulated first quarter 1991 personal income or GDP estimates for political purposes. BEA generally followed its standard procedures for using employment data in these estimates and deviated from these procedures only when required by what we believe were reasonable technical judgments" (GAO, 1993, p. 1).

3.5 Current Problems and Activities

BEA's regular publication schedules are carefully developed to take into account the needs of users, balanced against the responsibility to produce data of high quality. In general, the four-month lag of the State quarterly and preliminary annual State personal income data and the eight-month lag in the release of more detailed annual State personal income estimates are timely enough for most purposes and cause few hardships for the users of these series.

For county and metropolitan area data, the fifteen-month lag required to produce these estimates is considered too long for many purposes and has limited the usefulness of these data. In an effort to address the issue of the timeliness of its local area estimates, BEA has recently been testing the feasibility of developing preliminary annual estimates of personal income for metropolitan areas and non-metropolitan portions of States. These estimates would be available with a seven-month lag.

3.6 Conclusions

Rapid advances in computer technologies continue to provide improvements in the range of regional data available for local area estimation as well as in the timing of their availability. For example, the more timely availability of ES-202 wage and salary data coupled with BEA's improved computing capabilities and estimating procedures may allow for the much more timely release of preliminary income estimates for metropolitan areas.

These rapid advances in computer technologies also continue to expand the ease of data transfer, storage, and manipulation. For example, BEA recently introduced a CD-ROM containing the local area personal income estimates; many data users can now acquire the entire set of estimates rather than placing an order each time they need some of the data. As in the past, it is anticipated that these advances in electronic capabilities will continue to expand the uses and users of BEA's regional estimates.

REFERENCES

Advisory Commission on Intergovernmental Relations (ACIR) (1990), Significant Features of Fiscal Federalism, Volume 1: Budget Processes and Tax Systems, M-169, pp. 10-13, Washington, DC: U.S. Government Printing Office.

Bureau of the Census, U.S. Department of Commerce (1992), Statistical Abstract of the United States: 1992, Appendix II, Washington, DC: U.S. Government Printing Office.

Bureau of Economic Analysis (BEA), U.S. Department of Commerce (1985), Experimental Estimates of Gross State Product by Industry, BEA Staff Paper 42, Washington, DC: National Technical Information Service.

_____(1989), State Personal Income: 1929-87, Estimates and a Statement of Sources and Methods, Washington, DC: U.S. Government Printing Office.

_____(1991), Local Area Personal Income, 1984-89, Volume 1: Summary, Washington, DC: U.S. Government Printing Office.

Brown, R.L. and Stehle, J.E. (1990), "Evaluation of the State Personal Income Estimates," pp. 20-29, Survey of Current Business 70 (December, 1990).

Creamer, D. and Merwin, C. (1942), "State distribution of Income Payments, 1929-41," Survey of Current Business 22 (July 1942).

Nathan, R.R. and Martin, J.L. (1939), "State Income Payments, 1929-37," mimeographed report, Washington, DC: Bureau of Foreign and Domestic Commerce, U.S. Department of Commerce.

Parker, R.P. (1984), "Improved Adjustments for Misreporting of Tax Return Information Used to Estimate the National Income and Product Accounts, 1977," pp. 17-25, Survey of Current Business 64 (June 1984).

Trott, E.A., Dunbar, A.E., and Friedenberg, H.L. (1991), "Gross State Product by Industry, 1977-89," pp. 43-59, Survey of Current Business 71 (December 1991).

U.S. General Accounting Office (GAO) (1993), Gross Domestic Product: No Evidence of Manipulation in First Quarter 1991 Estimates, Washington, DC: U.S. Government Printing Office.

Zabronsky, D. (1988), "Reliability of the Census Journey-to-Work data in the Residence Adjustment for County Personal Income," Discussion Paper #35, Bureau of Economic Analysis, U.S. Department of Commerce.

CHAPTER 4

Postcensal Population Estimates:
States, Counties, and Places

John F. Long, U. S. Bureau of the Census

4.1 Introduction and Program History

The U. S. Bureau of the Census produces population estimates for the nation, states, counties, and places (cities, towns, and townships) as part of its program to quantify changes in population size and distribution since the last census. These estimates provide updates to the population counts by demographic and geographic characteristics from the last census. They also indicate the pace of population change since the last census and the relative influence of the components of population change. While the national estimates can be produced by a careful accounting system that adds annual births, deaths, and international migration to the previous year's population, subnational estimates require development of methods for dealing with the largely unmeasured component effects of internal migration. Many of these methods represent the type of *small domain* estimates that constitute the subject of this report.

4.1.1 Uses of postcensal population estimates

There are five major categories of uses for the Census Bureau's population estimates: 1) Federal and state funds allocation, 2) denominators for vital rates and per capita time series, 3) survey controls, 4) administrative planning and marketing guidance, and 5) descriptive and analytical studies. These five uses vary depending on whether the use is for a national, state, county, or place purpose.

National: - Survey Controls
          - National Social and Economic Series
          - Descriptive and Analytical Studies
          - Controls for Subnational Estimates

State: - Direct Federal Fund Allocation Formulas
          - Indirect Federal Fund Allocation

- Denominators for Federal and Other Data Series
- Federal Regulatory Actions
- Survey Controls
- Descriptive and Analytical Studies
- Controls for Substate Estimates

Counties: - Fund Allocation by State Governments
- Denominators for Federal, State, and Other Data Series
- Regulatory Action by State Governments
- Guides for Government and Private Sector Planning
- Descriptive and Analytical Studies
- Federal Data Series
- Controls for Subcounty Estimates

Places: - Federal Block Grants
- Fund Allocation and Regulatory Actions by Federal and
    State Governments
- Descriptive and Analytical Studies
- Government and Private Sector Planning
- Private Sector Marketing Efforts
- Base Data for Private Sector Data Development

More than 70 federal programs distribute tens of billions of dollars annually on the basis of population estimates (GAO, 1990). Even more money was distributed indirectly on the basis of indicators which used population estimates for denominators or controls (GAO, 1991). Many states also use the postcensal subnational estimates to allocate state funds to counties, townships, and incorporated places within the state.

A large number of Federal statistical series including state and county per capita income, national and state birth and death rates, and county level cancer rates by age, sex, and race use the results of the postcensal estimates. While many Federal agencies directly collect time series data on events and amounts, they require annual postcensal estimates of state and county population to produce per capita rates. These series provide an indication of national and subnational trends for fertility and mortality rates, incidence of cancer and other diseases, per capita economic changes, and other social, demographic, and administrative indicators.

Population surveys require independent controls from national population estimates by age, sex, race, and ethnicity as well as data on the geographic distribution of the population by states and selected metropolitan areas. These estimates are used to weight the sample cases such that the survey results equal the postcensal estimates used as controls. Each of the major surveys conducted by the Census Bureau control to somewhat different levels of geographic and demographic detail (Table 4.1). There are a number of reasons to control surveys to independent estimates. They were initially instituted to reduce the variance of the survey estimates. They are also used for a number of secondary reasons: reduction in month-to-month variability of longitudinal data from consecutive surveys, partial correction for the large rates of undercoverage of surveys relative to the census, and improved consistency between different surveys and other population data series based on independent estimates.

Table 4.1. Major Population Surveys Controlled To National, State And Substate Postcensal Population Estimates

Survey	Items Controlled
Current Population Survey	Monthly civilian noninstitutional population, by age, sex, and race (white, black, other) and Hispanic for the nation and total population 16+ for states, New York City and Los Angeles.
Survey of Income and Program Participation	Controls derived from CPS processing.
Consumer Expenditure Survey	Monthly total population by age, sex and race for the nation.
Hospital Discharge Survey	Annual civilian population by age and sex for the nation and regions.
National Crime Victimization Survey	Annual civilian population 12+ for the nation and 11 states.
National Health Interview Survey	Civilian noninstitutional population under 18, 18-44, 45-64, 65+ and total by sex quarterly for the nation and annually for states.
National Hunting and Fishing and Wildlife Associated Recreation Survey	Noninstitutional, nonbarracks population by age, race and sex for the nation and states for calendar years ending in 0 and 5.

There are numerous other administrative and analytical uses of the postcensal population estimates. They provide the only regular mechanism by which the components of population change are combined to track changes in the size and demographic and geographic distribution of the nation's population. The postcensal estimates provide essential information for administration and planning in the government and private sectors. In addition, they are used as a standard by state and local governments and the private sector in producing their own population estimates for smaller scale geography or for greater social and economic detail.

4.1.2 History of Census Bureau estimates program

Since the early 1900s, the Census Bureau has produced national population estimates. The methodology for these estimates developed into a component method in which the measured components of population change (births, deaths, immigration, and emigration) are added to or, in the case of deaths, subtracted from the most recent decennial census to estimate the current population.

When the Census Bureau attempted state population estimates beginning in the 1940s, it faced the difficult prospect of adding internal migration to the other components of population change. Since annual measures of internal migration by state are not available, many attempts were made to develop other ways to estimate state population change.

Through 1960, the principal method (known as *Component Method II*) was to estimate net migration based on annual changes in school enrollment. In the 1960s, a second method was added that estimated changes in the population level rather than measuring the components of population change. This method (the *ratio-correlation method*) uses regression analysis that relates changes in selected independent variables to changes in state population since the last census. These independent variables come from federal or state data sources. In the 1960s, the major proxy variables were vital events, school enrollment, tax returns, number of votes cast, motor vehicle registrations, and building permits. In the 1970s, the variables for votes cast and building permits were dropped and a variable for the size of the work force was added.

As the demand for estimates spread to the county level, the Federal State Cooperative Program for Population Estimates was formed to involve state governments in a joint effort with the Census Bureau. This organization permitted the extension of Component Method II, the ratio correlation method, and a housing unit method to the county level by providing data on school enrollment and various state administrative data systems at the county level. This system permitted the flexibility of using data sets selected for each individual state.

The enactment of General Revenue Sharing created a demand for population estimates for all general purpose governments(incorporated places, towns, and townships). To estimate these subcounty areas, the Census Bureau returned to a component based method (the *administrative record method*) in which migration was estimated using income tax data from the Internal Revenue Service (IRS). This method required matching addresses on successive years of tax returns and calculating a migration rate based on the total number of exemptions that moved into and out of each area. The key challenge in developing this methodology was to design a suitable method of coding mailing addresses to counties, incorporated places and minor civil divisions. The result was a probability coding guide based on a question on place of residence placed on the tax returns in selected years. This methodology proved so successful that it was added as an independent method in the estimation of state and county populations as well.

4.2 Program Description, Policies, and Practices

The level of geographic and characteristic detail and the methodologies of the current population estimates program are legacies of the expansion of estimates demands during the last three decades. Tables 4.2 and 4.3 show the frequency, detail, and methodology used at each geographic level of the Census Bureau's population estimates program.

Table 4.2. Postcensal Population Estimates Program for the 1990s

Area	Frequency (Base date)	Estimated Characteristics
Nation	Monthly	Total Population
Nation	Yearly	Age, Sex, Race, Hispanic
States	Yearly	Total Population, Age, Sex
Counties & Metropolitan Areas	Yearly	Total Population
Incorporated Places and Minor Civil Divisions	Even Years	Total Population

While the national population is estimated by age, sex, race, and Hispanic Origin, the subnational population estimates vary greatly in demographic and socio-economic detail. In general, the level of characteristic detail declines as the level of geography becomes finer.

Each level of geography also has its own combination of methods and input data. State population is currently produced on an annual basis by age and sex. County estimates are produced annually for the total population and, on an experimental basis, by age, race, and sex. Estimates for the total population of metropolitan areas are produced annually by summing the appropriate county data and by making adjustments for New England areas which are composed of townships rather than counties. Every other year, the Census Bureau produces total population estimates for incorporated places, towns, and townships.

Table 4.3. Estimation Method by Geographic Category: 1980-1989

	Component Methods			Change in Stock Methods				Mixed Methods	
	Cohort	Adm Rec	Meth II	Ratio Corre- lation	Hous- ing Unit	Group Quar- ters	Medi- care Enroll- ment	Com- posite	Com- bined
Nation	X								
State		X				X	X	X	X
County		X	X	X	X	X	X		X
Places		X				X			

4.3 Estimator Documentation

The methodology for postcensal estimates varies by level of geography with the widest array of methods used in county estimates. This methodological discussion focusses on the county estimates with occasional extensions to include methods specific to states or places. Postcensal population estimates update the last census population based on changes in the population or in components of population change. Actual information on such components of population change as births and deaths or on changes in symptomatic indicators related to changes in the population since the last census provide benchmarks to anchor the estimates.

The art of postcensal estimation of population comes in choosing appropriate benchmarks (or auxiliary data) to use in estimating the population change since the last census. One type of benchmark data, population flow data, consists of measures of the components of population change (eg. births, deaths, internal and external migration). The other type of benchmark data, population <u>stock</u> data, includes indicators that are correlated with population size and uses changes in those indicators to estimate the total change in population. Methods based on each of these two classes of data are found in several variations in the Census Bureau's postcensal population estimates program.

4.3.1 Flow methods

Flow methods are also known as component methods. They require some estimation of each of the components of population change since the last census. In the most general form, the *component method* reduces to a basic accounting equation for population change.

$$P_{i,t} = P_{i,0} + (B_i - D_i + I_i - E_i) + \mu_i P_{i,o} \qquad (4.1)$$

where,

$P_{i,t}$ = population estimate for area i at time t,
$P_{i,0}$ = population in area i at beginning of period,
B_i = births in area i since beginning of period,
D_i = deaths in area i since beginning of period,
I_i = international immigrants to i,
E_i = international emigrants from i, and
μ_i = estimator of rate of net internal migration to i.

Direct measurement is possible for some of the components of population change -- births, deaths, and some components of international migration. Vital statistics registration data do a good job of measuring the effects of natural increase but problems can arise with immigration data. Unmeasured migration across international borders is a major cause of estimation error that requires the use of assumptions about the quantity and characteristics of the population flows missed. The issue of small area estimation as defined by this report arises when there is no direct measure of the component of interest. In equation 4.1, the rate of internal net migration (μ_i) is not directly measured but must be estimated from an alternative data source devised for another purpose.

In order to simplify the task of finding a good estimator of net internal migration, we confine the use of equation 4.1 to the household population under 65. The population 65 and over and the population in group quarters (military barracks, prisons, college dormitories, etc.) have different patterns of migration and are better handled separately later in the process by looking specifically at data systems that explicitly measure changes in the 65 and over population and in the group quarters population.

One method for estimating the net internal migration rate (μ_i) for the household population under 65 uses administrative data that provide addresses for individuals at two different points in time (usually a year apart). Such data provide approximate data on inmigration, outmigration, and even area-to-area flows. While there are several potential sources of these administrative data -- changes in postal addresses, drivers license records, tax returns, and health insurance information -- the problem is to find a source that provides representative coverage and consistency in reporting and tabulation. The Census Bureau uses an *administrative records method* that compares tax returns from the Internal Revenue Service (IRS) for changes in filing addresses between two consecutive annual tax filings (U. S. Bureau of the Census, 1988). In the estimates process, tax returns from one year are matched with those from previous years by matching Social Security numbers of the filers. For persons with a new address, the new mailing address is coded to state, place, and county. If the state, place, or county is different from the previous year, the filer and all exemptions are classified as migrants. These data are then used to construct net migration rates for each county and place as an input to the population estimation formula. An estimate of the rate of net migration is calculated by dividing the net flow of exemptions (the tax filer plus his or her dependents) moving into the area by the number of exemptions filed in the area (See equation 4.2).

$$\mu_i = \frac{\Sigma_j (T_{ji} - T_{ij})}{T_{i.}} \tag{4.2}$$

where,

> $T_{i,j}$ = flow of tax exemptions from area i to j,
> $T_{i.}$ = total number of matched tax exemptions living in area i
>   at the beginning of the period.

This net migration rate is then multiplied by the initial population as shown in equation 4.1. A critical assumption in this method is that the population not covered by the administrative data set moves similarly to the population covered or that the uncovered population is too small to affect the results markedly. Since this assumption is especially inappropriate for the population over 65 and for certain military and institutionalized populations, those populations are handled separately as explained below. Other potential problems include the difficulty of coding addresses to geography, changes in administrative coverage over time, and the elimination of administrative data sources as governmental programs change.

A second method of estimating the net internal migration rate (μ_i) uses school enrollment data. Changes in the size of the population enrolled in elementary and secondary school can be used to estimate the net migration of the general population. In one such method, *component method II*, changes in school enrollment are compared to expected changes due to natural increase alone in order to produce indirect estimates of the *net* migration rate of the school-aged population (Batutis, 1991). The migration rate for the total population is then estimated by adding the difference between the net migration rate of the total population and the net migration rate of the school-aged population in the most recent census. The critical assumption here is that the relationship of net school-aged migration and net total migration remains constant over time.

4.3.2 Change in Stock Methods

A fundamentally different approach to population estimates emphasizes the total change in population size since the last census rather than demographic components of change. These *change in stock* methods relate changes in population size to changes in other measured variables that are assumed to be correlated with population change.

The choice of possible variables is wide: number of housing units, automobile registrations, total number of deaths (and or births), tax returns, etc. Note that births and deaths in this method are not viewed as components but as indicators of the size of the population. Similarly, drivers licenses and tax returns are not used as indicators of migration as they were in the flow methods but as proxies for the size of the total population.

The U. S. Census Bureau county estimates program uses a special case of the change in stock method known as the *ratio-correlation method* (Namboodiri, 1972). In this method, we construct a linear regression equation for each state separately, using indicators appropriate for that state. The independent variables are ratios of the proportion of each indicator that is located in a given county in the state as of the date of the most recent census to the comparable proportion at the time of the prior census. The dependent variable is the ratio of the proportion of a state's population in a given county in the most recent census to the comparable proportion in the prior census. The resulting regression parameters(k,α,β,γ) are then used to estimate postcensal county populations in equation 4.3.

$$P_{i,t} = P_{s,t} \frac{P_{i,0}}{P_{s,0}} [k + \alpha \frac{A_{i,t}/A_{s,t}}{A_{i,0}/A_{s,0}} + \beta \frac{B_{i,t}/B_{s,t}}{B_{i,0}/B_{s,0}} + \gamma \frac{C_{i,t}/C_{s,t}}{C_{i,0}/C_{s,0}}] \qquad (4.3)$$

where,

$P_{s,0}$ = state population in the last census,
$P_{s,t}$ = independent estimate of current state population,
$A_{s,0},B_{s,0},C_{s,0}$ = indicator variables for state total at date of last census,
$A_{i,0},B_{i,0},C_{i,0}$ = indicator variables for county i at date of last census,
$A_{s,t},B_{s,t},C_{s,t}$ = indicator variables for state total for estimate date,
$A_{i,t},B_{i,t},C_{i,t}$ = indicator variables for county i for estimate date.

The key assumption in this method is that the relationship among geographic units between change in population and change in the selected indicator variables remains constant over time (Tayman and Schafer, 1985). Complications also arise if indicator variables change over time in selected areas for reasons unrelated to population -- for example, changes in the tax law, changes in general fertility rates, increases in automobile registrations per person, etc. Another population stock method used to estimate the ratio of the current population to the household change is the *housing unit method*. In this method, tax rolls, construction permits, certificates of occupancy, or utility data could be used to calculate changes in the number of housing units in an area (Smith and Mandell, 1984). In the Census Bureau's methodology the housing stock from the last census is updated using data on housing construction, demolitions, and conversions (Eq. 4.4).

$$U_{i,t} = (U_{i,0} + V_i - W_i)$$
(4.4)

where,

$U_{i,0}$ = housing units in area i in the last census,
$U_{i,t}$ = estimated housing units in area i for estimate date (t),
V_i = housing units constructed in area i since last census,
W_i = housing units in area i demolished since the last census,

The number of households in area i for date t is estimated by multiplying the estimated number of housing units at time t by an updated estimate of the occupancy rate for area i at time t. By assuming that the local occupancy rate changes as the national rate, we can update the area's rate by multiplying the occupancy rate for area i at the time of the census by the ratio of the national occupancy rate at time t from the Current Population Survey (CPS) to the national occupancy rate at the time of the census.

$$H_{i,t} = U_{i,t} \frac{H_{i,0}}{U_{i,0}} \frac{H_{.,t}/U_{.,t}}{H_{.,0}/U_{.,0}}$$
(4.5)

where,
   $U_{.,0}$ = national housing units in the last census,
   $U_{.,t}$ = national housing units for estimate date,
   $H_{i,0}$ = households in area i in the last census,
   $H_{i,t}$ = households in area i for estimate date,
   $H_{.,0}$ = national households in the last census,
   $H_{.,t}$ = national households for estimate date.

Finally, the population for the area i is calculated by multiplying the area's household estimate by an updated estimate of population per household. Again we assume that the area's population per household from the last census can be updated by multiplying by the ratio of the national population per household from the CPS to the national population per household in the last census.

$$P_{i,t} = H_{i,t} \frac{P_{i,0}}{H_{i,0}} \frac{P_{.,t}/H_{.,t}}{P_{.,0}/H_{.,0}}$$ (4.6)

where,
   $P_{i,t}$ = estimated population in area i,
   $P_{.,0}$ = national population in last census,
   $P_{.,t}$ = national population at the date of the estimate.

All of the methods discussed so far refer to the household population under 65. The two other segments of the population, the population 65 and over and the group quarters population, are measured by their own specific change in stock methodologies. Since these two groups have unique characteristics (especially in terms of their migration patterns), we use administrative records systems that are unique to each of the two groups. The population over 65 is estimated by using changes in the medicare population since the last census as a direct measure of the change in the population 65 and over. No such nationwide systems exists for the group quarters populations (defined for estimates purposes as the population in military barracks, college dormitories, prisons and other institutions). Changes in these population since the last census are obtained from an inventory of major group quarters locations that is maintained and annually updated by a special data collection process in the Population Estimates Branch of the Population Division in cooperation with state agencies affiliated with the Federal-State Cooperative Program for Population Estimates.

4.3.3 Combined methods

The U. S. Census Bureau's postcensal population estimates program combines methods in two ways. Within each level of geography (states, counties, and places) several of the above methods are combined (Table 4.3 above). Since certain methods represent given subpopulations better, a combination of methods may be viewed as more robust -- less likely to change due to extraneous factors that might affect one or the other of the estimates. There is a further mixing of methods since the estimates at each level of geography are controlled to the results of the estimates made at the next higher level of geography.

The methodology for making state estimates during the 1980s averaged the results of the administrative record method with those of the *composite method*. In the composite method, the population is divided into three age groups, each of which is estimated by a separate method. The population under 15 is estimated using changes in the levels of school enrollment (similar to Component Method II). The population ages 15-64 is estimated by a ratio- correlation method in which the independent variables are tax returns, school enrollment, and housing units. The population over 65 is estimated using a method in which changes in the number of persons on medicare since the last census date are added to the population aged 65 and over at the last census (U. S. Bureau of the Census, 1984). The total state population by age is then controlled to equal the estimated national population age structure.

Annual county population estimates are produced independently for each state to coincide with the state's total population estimated above. A distinct methodology for each state is developed in consultation with that state's member of the Federal-State Cooperative Program for Population Estimates. In most cases, it consists of the average of two or three of the methods described above: the administrative records method, component method II, and the ratio-correlation method. Moreover, within the ratio-correlation method, different states use different independent variables which may include school enrollment, tax returns, medicare enrollment, automobile registrations, births, deaths, dummy variables for county size, or other state-specific data series. Additional adjustments are made for changes in selected military and institutional populations and for changes in the population over 65. Final results are controlled to the state population estimate produced by the Census Bureau using a uniform method across all states (van der Vate, 1988).

Place estimates use a strict administrative record methodology where migration is based solely on the migration rates derived from changes in addresses on tax returns. The only other adjustments for place estimates are for changes in selected military and institutional populations and a final control to county level population estimates (U. S. Bureau of the Census, 1980).

4.4 Evaluation Practices

The estimation process demands continuous vigilance. Methods that appear to work well at the beginning of a decade may be unsatisfactory later in the decade. Only constant testing, data evaluation, quality control, and checks for reasonableness can ensure a sound program of population estimation.

Whatever the method of estimation chosen, a number of considerations should be kept in mind. No matter how sophisticated the methodology, the estimate will only be accurate if the underlying assumptions hold and the input data are reliable. Many things can happen to endanger these conditions. For example, the relationships that held between variables in a previous decade might no longer hold in the current decade. The data series that one is depending upon to update the population may deteriorate or fail to measure the same underlying phenomenon as conditions change. Even if the administrative or other indicator data measure the population well, there may well be problems of geographic coding that fail to assign the population to the correct geography.

Finding an appropriate yardstick against which to measure the postcensal population estimates is difficult. During the decade, aside from special censuses for a handful of places, there are no suitable numbers to compare to the estimates -- thus we know little about the short run accuracy of population estimates. We can only measure their accuracy at the extreme end of their range (after 10 years) using the next decennial census. Even here, the changing level of coverage between censuses for any given area can lead to imprecision in our measurement of estimates accuracy. Using the results of the 1980 and 1990 censuses as enumerated, the Census Bureau evaluated the accuracy of the population estimates program. The results (summarized in Table 4.4) show that population estimates made for the nation, for states, and for counties were reasonably accurate, but that estimates made for small places were quite inaccurate. Estimates for places under 5,000 had a mean absolute error of more than 15 percent while places over 50,000 had a mean absolute error of less than 5 percent.

The last two columns in Table 4.4 present a more telling comparison. Column two compares the 1990 census and the provisional 1990 postcensal estimate while column three compares the 1990 census with the 1980 census. For most levels of geography the postcensal population estimate provides a far more accurate estimate than simply holding the population constant at the level of the last census. For example, state postcensal estimates had an mean absolute error of only 1.5 percent, while holding the last census constant would give an error of 10.0 percent.

Table 4.4. Estimates of Accuracy (Mean Absolute Percent Error) by Level of Geography: 1970-80 and 1980-90.

Unit/Size	1980 Estimates vs. 1980 Census	1990 Estimates vs. 1990 Census	1980 Census vs. 1990 Census
Nation	2.1	0.5	9.5
States	2.5	1.5	10.0
Counties	4.2	4.0	10.0
Places: Over 50,000	4.3	4.0 *	10.8
Places: 5,000 to 50,000	7.2	7.1 *	11.7
Places: Under 5,000	16.6	19.0 *	18.1

* Place estimates for 1990 are provisional (based on extrapolations of 1988 estimates).

Source: Population Division, U. S. Bureau of the Census.

On average, the estimates methodology is also much better than using the last census for counties and incorporated places over 5,000 population. However, for many incorporated places under 5,000, holding the population constant at the 1980 level would have given more accurate results that did our postcensal estimate program.

These inaccuracies for small places may be due to a number of sources: The problem of coding administrative records to small units of political geography, the greater importance of migration in population change for small areas, and the greater likelihood that the broad assumptions that might apply on average for larger areas would not apply to small localities with very specific characteristics. Since the Census Bureau is required by law to produce data for all incorporated places and townships, we will need to show places under 5,000 as well as the larger places for which we can produce good estimates. However, it is incumbent on us to show the uncertainty in the estimates for small areas in future publications in addition to making continual progress in refining and improving our estimates methodologies and data bases.

4.5 Current Problems and Planned Activities

Many of the problems of the current population estimates system are the results of its past success and rapid growth during the 1960s and 1970s. Each new program, each expansion of characteristic detail, each reduction in the size of geographic unit has been accompanied by new data sets, by new methods, and by new production procedures. Although the Census Bureau has done a good job of meeting users expectations as these demands have increased, there is room for improvement in the estimates methodology and operations.

We have embarked on a set of seven initiatives to revamp the population estimates program and lead it into the next century. These initiatives fall under the following headings: 1) defining the mission, 2) methodological integration, 3) input data quality, 4) geographic flexibility, 5) characteristic detail, 6) analysis of trends, and 7) production efficiency.

4.5.1 Defining the Mission

The products currently estimated by the Census Bureau's Population Estimates Program are the results of opportunities and legislative requirements over a period of three decades. We plan to reexamine the demands for and uses of population estimates. A thorough study of the needs for population estimates and the Census Bureau's proper mission in filling those needs is an initial priority. We are currently polling a number of our users -- Federal government agencies, the Federal-State Cooperative Program members, private data vendors, and a number of other groups to ascertain their needs for population estimates.

Some of the suggestions received so far involve modifying the population estimates program in order to produce more detailed characteristic information at the state and county level. We hope to produce age, sex, race, and Hispanic Origin data for counties. With more research, we may also be able to produce the county-level data on households -- number, size, and income -- that is currently demanded by many users. We are examining the feasibility of producing estimates for larger places on a yearly basis and producing estimates for other subcounty geography as well -- possibilities include census tract aggregates, subareas within large cities, and (for some purposes) Zip codes.

4.5.2 Methodological Integration

The many different methods of estimating population developed over the past decades have resulted in a complex population estimates program. The need now is to integrate these disparate methods into an orderly system. Traditionally, the various estimation models used at the Bureau have been integrated by a simple averaging of the different estimates at a given level of geography and by controlling the sum of estimates at one level of geography to the averaged estimate at the next higher level.

The time has come to reexamine each set of methods for suitability as parts of an integrated, parsimonious model for producing population estimates. In order to discuss methods of integrating our current methods, it is useful to distinguish between methods that measure the changes in the population stock and those that measure the components of population change. Methods showing the change in population stock (the ratio correlation method, the medicare change methodology, and the change in group quarters population) use changes in proxy variables since the last census to produce estimates of the total net change since the last census. These methods permit the use of many symptomatic measures of population size that may not be amenable to a flow approach.

Component methods such as the administrative records method and component method II represent flow methods in which the components of population change births, deaths, international migration, and internal migration are each measured separately and added to or subtracted from the initial population. The advantage of this type of method is that it gives an estimate ont only of the population but also of the components of population change. This method provides additional information about the reasons for change, the reasonableness of the estimates, and provides inputs for population projections. Component methods are often preferable for larger areas because they use relatively accurate counts of births and deaths to compute a large part of population change. Consequently, administrative records which are often less accurate need only be used to estimate the portion of population change due to migration. Current research at the Census Bureau is underway to quantify the relative effects of errors in each component on the final population estimates. For small area, these advantages disappear and change in stock methods such as the housing unit method may be more appropriate.

As we integrate methods, we should be careful to retain the flexibility offered by multiple independent methods of estimating population. Since methodologies for population estimates are dependent upon the use of data sets collected for purposes other than population estimates, the quality and availability of a given input data set is never certain. Only with multiple methods can we be assured of the ability to produce population timely and reliable population estimates. Multiple methods also provide a necessary check on the validity of the estimates results; surprising changes in demographic trends can be checked using independent sources in order to see if the results are merely idiosyncracies of a given input data source. The existence of independent methods of estimating population could prove a distinct advantage in trying to gauge the accuracy of estimates between censuses. We should examine the potential of using measures of divergence between independent estimates to determine the reliability and degree of confidence we have in the accuracy of postcensal estimates. If three independent estimates give very close values, we should have more confidence in those estimates than if the estimates vary widely.

4.5.3 Input Data Quality

Perhaps even more important than the type of method chosen is the choice of data set used in the estimate. Producing postcensal population estimates requires integrating traditional demographic data sets such as census results, birth and death records, and immigration statistics with nontraditional sources collected for other administrative purposes such as tax returns, school enrollment, drivers' licenses, housing construction, survey data, etc. The art of population estimation is to combine these traditional and nontraditional sources to make maximum advantage of all the data available.

The most challenging aspect of working with population estimates is the use of data sets designed and collected for administrative purposes rather than for statistical or demographic purposes. Ideally, such data sets should have universal coverage, change in direct relation with population changes, and be consistent over time in content and form. No data set actually meets these criteria. The level of population coverage is often less than 100 per cent. Programmatic changes or changes in social behavior independent of population change may affect the coverage rate. Worst of all, the administrative data set may even disappear if its programmatic need or funding disappears.

Consequently, a healthy population estimates program requires careful attention to the quality and timeliness of input data as well as to the reliability of access to the input data. This requires working with our data providers to monitor the input databases on a number of requirements including reliability, consistency, coverage, characteristic detail, and idiosyncracies produced by programmatic and other changes. It also entails work with data producers to address questions of mutual interest such as cost, confidentiality, and legal requirements for data handling. Since administrative datasets may disappear over time, work must also continue on nurturing alternative data sets to provide similar or superior data. The need for flexibility to address changing data set availability and quality is yet another argument for using multiple independent methods and data sets to provide redundancy in the estimates program.

4.5.4 Geographic Flexibility

Linking data on population to geography is the key to population estimation methodology. Any system for making subnational population estimates must have a credible method for developing such geographic correspondence. Population estimates are required for legally defined geographic entities such as counties and incorporated places and the estimates methodology must take these requirements into account. In the county estimates conducted jointly with states under the Federal-State Cooperative Program, we assume that the input data used in the ratio correlation methodology systems are correctly coded by county of residence.

In the administrative records method matching tax returns to determine state, county, and place migration, the Census Bureau must provide the geographic coding for movers based on the mailing addresses of filers from the IRS tax forms. In order to categorize these filers by county and place of residence, the current methodology uses a probability coding guide. With the aid of data from a residence question on the 1980 tax form, mailing addresses were categorized by P.O. name, state, zip code, and address type (street address, P. O. Box, RFD route) and assigned a probability of falling within each of 3100 counties and 39000 places.

There are several problems that lead to deterioration in the coding guide over time. Some of the more obvious ones can be corrected by manual adjustments in the coding guide, eg. creation of new Zip codes or revised boundaries for old Zip codes, changes in the boundaries of incorporated places, etc. A key cause of deterioration that cannot be fixed is the change over time in the distribution of the population within a given address key (post office, state, Zip code, address type combination). To the extent that those changes in distribution cross county, town, and city boundaries, the resulting coding will be incorrect. Moreover, the probability system itself may well put individual persons in the wrong county or place. We know little about how these errors propagate through the system after several years and multiple migrations.

The Census Bureau is currently developing a new geographic coding system that permits frequent updating and, if possible, exact matching rather than probability matching of addresses to geography. The system is based on the "master address list" proposed by the Census Bureau's Geography Division as an outgrowth of the development of the TIGER digitized mapping project and the "address control file" created for use with the 1990 census. This system would provide an annually updated digitized data base that could place most addresses in the United States into the appropriate census block (and thus into any unit of geography that also has its boundaries in the TIGER system). In the estimates area, we are exploring the feasibility of developing a coding system that would code street addresses to subcounty areas using such a master address list. The existence of a continuously updated master address list could provide far greater geographic detail, ease of updating and correcting for boundary changes, and flexibility in dealing with changing geographic concepts and shifts in population distribution.

This methodology also provides the promise of a far greater benefit in the future. The ability to provide exact matching based on geography might one day permit the matching of records on the basis of address rather than an identifier such as social security number. Such an ability would provide the opportunity to bring far more information sources to bear on the estimation effort.

4.5.5 Characteristic Detail

Another major area for innovation is the expansion of data on population characteristics -- both demographic characteristics such as age, race, and sex and social/economic characteristics such as household structure and income. In order to get a better hold on the demographic structure of substate areas and to use as a denominator in calculating incidence rates, there is a major increase in the demand for age, race, and sex distributions at the county level between censuses. These data are not available from the IRS tax records that form the principal part of our administrative records processing. Consequently, we are developing alternative methods to provide these data for counties and large places as an integral part of the estimation process.

We have experimented with a number of possible approaches. One of these experimental programs developed a projected estimate by which county trends in migration by age, race, and sex from the previous decennial census were extrapolated into the current decade, added to actual birth and death rates to produce a population by age, race, and sex that was then controlled to the official estimate of total population for a county. Another experimental program extends the current administrative record method by adding information by age, race, and sex from Social Security records to a sample of IRS returns to provide internal migration data for states and large metropolitan areas. Current plans call for integrating these programs into our standard procedures by the mid 1990s.

There are also possibilities for using survey data combined with administrative data to obtain characteristic information. While matched survey and administrative data records on an individual basis may prove difficult, there have been efforts to combine data on an aggregate basis. A recent example is an analysis of internal migration that combined aggregate data from the decennial census, matched tax return migration data, and survey data from the Current Population Survey (CPS) to provide a time series of migration by characteristics for state to state flows. Research is proceeding on whether more information from surveys could be combined with the administrative record methods by either aggregate or individual statistical modeling approaches.

Another major effort is underway to produce estimates for housing units and households for survey controls to the American Housing Survey and other housing based surveys. This program uses data on additions and deletions from housing stock to update the housing inventory from the decennial census. While this method is similar to the housing unit method for population estimates described above, the resulting housing unit estimates are used directly as survey controls rather than only used to estimate population.

There is also the potential for integrating more administrative data into the estimates procedure. A number of federal, state, and even private data sets have been suggested. Possible data sets include state tax data, post office change of address forms, state drivers license information, food stamp enrollment information, utility hookup records, and telephone directory information. These and other data sets will be explored for their potential utility for making subnational estimates assuring proper attention is given to protection of privacy and proper disclosure safeguards.

4.5.6 Analysis of Trends

A prime advantage of the population estimates programs is its information on the changes in spatial population distribution between censuses. While the Census Bureau has put great emphasis on the production of estimates for individual states, counties, and places, we have only occasionally provided the summary information on the broader trends in population redistribution. An analysis of population redistribution trends between cities and suburbs, high and low density areas, areas of high and low unemployment, and other analytical categories should be an annual part of our activities. In order to do this, a simple first step is to classify counties by relevant analytical characteristics so that such summaries could be a standard part of our processing. In addition, we plan an annual analytical report on population distribution trends based on the entire range of population estimates.

Much of the intermediate data on components of population change (migration, births, deaths, numbers of housing units, etc) used in constructing the population estimates is of analytical interest in its own right. These data should be developed as their own data products and used to provide an analytical view of the dynamics of current population change. An integrated set of historically consistent data series on births, deaths, international, and internal migration should be developed for all major geographic areas for which population estimates are produced. As a first step, we are producing a consistent time series of population counts for all counties and for cities over 25,000 from 1790 through 1990.

4.5.7 Production Efficiency

The uncoordinated and erratic growth pattern in the population estimates area has had a substantial effect on production efficiency. During the 1980s, delays in production and unreliable publication dates have frequently resulted from the unwieldiness of the current production process. For the 1990s, we have streamlined the production process as a result of more parsimonious methodologies and a more focused set of products. Many of our users repeatedly tell us that it is more important to have a firm production date than to be too optimistic in our timetables. Efforts toward redesigning the estimates product have as a major goal a firm production schedule with realistic deadlines. While considerable progress has been made on this commitment, we expect to strive toward continuous improvement in timeliness as well as reliability and cost reduction.

4.6 Conclusion

Postcensal population estimates are an integral part of the U. S. statistical system -- combining census results with tabulations on vital events, providing the population controls by which household survey results can be weighted, and producing a continuous and up-to-date time series of changing population size and distribution between censuses. These estimates are only possible with the creative use of censuses, vital events, administrative data, and other unconventional sources for estimating changes in population on a timely basis.

As we approach the twenty-first century, the population estimates program provides an ideal starting point for an integrated demographic and social accounting system. The system already unites the decennial census and population survey results through a series of longitudinal controls. These longitudinal controls are based on previous censuses and vital events, and could be modified to incorporate measurements of undercoverage if desired. In the 1990 census, the estimates system provided substantial information for coverage improvement during the operation of the census and in evaluating coverage after the results were in. The system provides the opportunity to integrate the results of administrative records collected for other purposes to augment and improve traditional demographic data. Our efforts to integrate our geographic coding with the decennial census data base (TIGER), to maintain estimates of housing units and households as well as population, and to use data on social and economic characteristics from surveys in the estimation process take us beyond a purely demographic system to an enhanced estimates program that could eventually provide continuously updated data on many of the variables now only measured by the census. Moreover, such an integrated estimates system could provide data on the components and rhythm of population, housing, geographic, social, and economic change that no individual data source can now provide.

REFERENCES

Batutis, Michael J. 1991. "Subnational Population Estimates Methods of the U. S. Bureau of the Census," U. S. Bureau of the Census, Population Division Working Paper.

General Accounting Office. 1990. *Federal Formula Programs: Outdated Population Data Used to Allocate Most Funds.* September. GAO/HRD-90-145.

General Accounting Office. 1991. *Formula Programs: Adjusted Census Data Would Redistribute Small Percentage of Funds to States.* November. GAO/GGD-92-12.

Mandell, M. and J. Tayman. 1982. "Measuring Temporal Stability in Regression Models of Population Estimation." *Demography*, 19:135-136.

Namboodiri, N. K. 1972. "On the Ratio-Correlation and Related Methods of Subnational Population Estimation." *Demography*. 9: 443-453.

National Academy of Sciences. 1980, *Estimating Population and Income of Small Areas.* Washington, D.C., National Academy Press.

O'Hare, W. P. 1976. "Report on a Multiple Regression Method for Making Population Estimate." *Demography*. 13:369-379.

O'Hare, W. P. 1980. "A Note on the Use of Regression Methods in Population Estimates." *Demography*. 17:341-343.

Roe, Linda K., John F. Carlson, and David A. Swanson. "A Variation of the Housing Unit Method for Estimating the Population of Small, Rural Areas: A Case Study of the Local Expert Procedure," *Survey Methodology*, 19: 155-163

Smith, Stanley K. and Bart Lewis. 1980. "Some New Techniques for Applying the Housing Unit Method of Local Population Estimation," *Demography*, 17: 323-339.

Smith, Stanley K. and Bart Lewis. 1983. "Some New Techniques for Applying the Housing Unit Method of Local Population Estimation: Further Evidence", *Demography*, 20: 407-413.

Smith, Stanley K. and Marylou Mandell. 1984. "A Comparison of Population Estimation Methods: Housing Unit Versus Component II, Ratio Correlation, and Administrative Records," *Journal of the American Statistical Association*, 79: 282-289.

Smith, Stanley K. 1986. "A Review and Evaluation of the Housing Unit Method of Population Estimation," *Journal of the American Statistical Association*, 82: 287-296.

Statistics Canada. *Population Estimation Methods: Canada.* Ottawa: Ministry of Supply and Services.

Swanson, David A. 1980. "Improving Accuracy in Multiple Regression Estimates of Population Using Principles from Causal Modelling," *Demography*. 17:413-427.

Swanson, David W. 1989. "Confidence Intervals for Postcensal Population Estimates: A Case Study for Local Areas," *Survey Methodology*. 15: 217-280.

Swanson, David W. and L. Tedrow. 1984. "Improving the Measurement of Temporal Change in Regression Models Used for County Population Estimates," *Demography*. 21: 373-381.

Tayman, Jeff and Edward Schafer. 1985. "The Impact of Coefficient Drift and Measurement Error on the Accuracy of Ratio-Correlation Population Estimates." *The Review of Regional Studies*. 15:3-10.

U. S. Bureau of the Census. 1980. "Population and Per Capita Money Income Estimates for Local Areas: Detailed Methodology and Evaluation," *Current Population Reports*. Series P-25, No 699.

U. S. Bureau of the Census. 1983. "Evaluation of Population Estimation Procedures for States, 1980: an Interim Report." *Current Population Reports*. Series P-25, No. 933.

U. S Bureau of the Census. 1984. "Estimates of the Population of States: 1970 to 1983," *Current Population Reports*. Series P-25, No. 957.

U. S. Bureau of the Census. 1985. "Evaluation of 1980 Subcounty Population Estimates," *Current Population Reports*. Series P-25, No. 963.

U. S. Bureau of the Census. 1986. "Evaluation of Population Estimation Procedures for Counties: 1980," *Current Population Reports*. Series P-25, No. 984.

U. S. Bureau of the Census. 1987. "State Population and Household Estimates, With Age, Sex, and Components of Change: 1981 - 1986", *Current Population Reports*. Series P-25, No. 1010.

U. S. Bureau of the Census. 1988. "Use of Federal Tax Returns in the Bureau of the Census' Population Estimates and Projections Program". Population Division Working Paper.

U. S. Bureau of the Census. 1988. "Methodology for Experimental County Population Estimates for the 1980s", *Current Population Reports*. Special Studies. Series P-23, No. 158.

U. S. Bureau of the Census. 1989. "Population Estimates by Race and Hispanic Origin for States, Metropolitan Areas, and Selected Counties: 1980 to 1985." *Current Population Reports.* Series P-25, No. 1040-RD-1.

U. S. Bureau of the Census. 1989. "County Population Estimates: July 1, 1988, 1987, and 1986," *Current Population Reports*, Series P-26, No. 88-A.

U. S. Bureau of the Census. "Population Estimates for Metropolitan Statistical Areas: July 1, 1988, 1987, and 1986," *Current Population Reports.* Series P-26, No. 88-B.

U. S. Bureau of the Census. 1990. "State Population and Household Estimates: July 1, 1989." *Current Population Reports.* Series P-25, No. 1058.

U. S. Bureau of the Census. 1990, "1988 Population and 1987 Per Capita Income Estimates for Counties and Incorporated Places," *Current Population Reports.* Series P-26, No. 88-SC.

van der Vate, Barbara J. 1988. "Methods Used in Estimating the Population of Substate Areas in the United States," Paper presented at the International Symposium on Small Area Statistics, New Orleans, LA, Aug. 26-27.

CHAPTER 5

Bureau of Labor Statistics' State and Local Area Estimates of Employment and Unemployment

Richard Tiller and Sharon Brown, Bureau of Labor Statistics
Alan Tupek, National Science Foundation

5.1 Introduction and Program History

The Bureau of Labor Statistics' (BLS) Local Area Unemployment Statistics (LAUS) Program produces state and area employment and unemployment estimates under a federal-state cooperative program. At present, monthly employment and unemployment estimates are prepared for the 50 states and the District of Columbia, all Metropolitan Statistical Areas (MSA's), all counties, and selected subcounty areas for which data are required by legislation -- more than 5,300 areas. The Current Population Survey (CPS), conducted by the Bureau of the Census for the BLS, is the official survey instrument for measuring the labor force in the United States. The CPS sample provides direct monthly survey estimates of employment and unemployment for the nation, selected states and New York City and Los Angeles. However, the CPS sample is not sufficiently large in most states and substate areas to provide reliable monthly estimates. Therefore, methods are used to combine data from other sources with current and historical CPS sample estimates to produce monthly estimates of employment and unemployment for the remaining states, the District of Columbia, and substate areas.

The CPS began during the Great Depression as a project of the Works Project Administration (WPA). During and following World War II, the need for unemployment data at the local level began to develop. A number of state and federal agencies began making estimates using various procedures. In 1950, the U.S. Department of Labor's Bureau of Employment Security, in an attempt to standardize the estimation methods, issued guidelines in a booklet: Techniques for

Estimating Unemployment. In 1960, the Handbook Method on Estimating Unemployment was issued. This building block or accounting method for developing total employment and unemployment estimates is essentially still used for substate areas today. About the same time, Congress began passing legislation using local unemployment data for the allocation of funds, such as the Area Redevelopment Act in 1961 and the Public Works Economic Development Act in 1965. Legislated programs which currently allocate funds to states and local areas based on unemployment estimates, include the "Disadvantaged Adults and Youths", "Summer Youth", and "Dislocated Workers" programs of the Job Training Partnership Act, the "Emergency Food and Shelter Program", and the "Public Works Program". In FY91, more than 9 billion dollars in appropriations to states and local areas were based, in full or in part, on local area unemployment statistics.

In 1972, the BLS acquired responsibility for unemployment statistics. BLS subsequently introduced changes to the Handbook methodology, including the use of annual average estimates from the CPS as controls for the state and area monthly estimates. Beginning in 1973, the CPS sample size was expanded to allow for direct sample based estimates for the 10 largest states, Los Angeles and New York City. In 1984, an 11th state was added. The Handbook method was still used for the 39 remaining states and the District of Columbia. However, a 6-month moving average adjustment using CPS data was applied to the state estimates. For substate areas, Handbook estimates are prepared for all labor market areas in the state, which are controlled to the state CPS-based estimates of employment and unemployment. At present, monthly employment and unemployment estimates are also prepared for all MSA's, all counties, and selected subcounty areas for which data are required by legislation. In 1989, a new methodology was introduced for producing monthly state employment and unemployment statistics for the 39 smaller states and the District of Columbia. This method is a time-series regression model, and uses a state-space Kalman filter approach.

Monthly estimates for the 39 smaller states and the District of Columbia are published approximately 6 weeks after the reference week of the CPS, which is the week including the 12th. Sample based estimates for the largest 11 states are released a few weeks earlier, (usually the first Friday of the month following the reference month) with the national estimates. Estimates for the 39 smaller states and the District of Columbia are revised a month later to reflect revisions in the Unemployment Insurance Statistics and the Current Employment Statistics (Payroll Employment Survey) Program, which are used in both the Handbook Method and State Modeling Method. At the end of the year, monthly state estimates are revised (benchmarked) so that their annual average equals the CPS sample based annual average. For the 11 large states, data are revised to reflect population controls.

5.2 Program Description, Policies, and Practices

Only five labor force estimates are published monthly for state and substate areas: Civilian noninstitutional population, civilian labor force, employed, unemployed, and the unemployment rate. Each month a press release - The Employment Situation - is issued and the Commissioner of Labor Statistics testifies before the Joint Economic Committee of Congress. The press release includes employment and unemployment estimates for the 11 largest states, in addition to national estimates. These data, as well as data for the remaining 39 states, the District of Columbia and Metropolitan Statistical Areas (MSAs) are published about four weeks later in Employment and Earnings. Seasonally adjusted data are provided for all 50 states and the District of Columbia, beginning in January 1992. Although the data for the smaller states are published with the data for the 11 largest states, the data are published in two sets in a table. The estimating methodology for the smaller states is provided in a footnote at the bottom of the page. A separate monthly publication - State and Metropolitan Unemployment - also includes data using direct sample survey estimates for the 11 largest states, the State Model methodology for the remaining states and the state CPS additively-adjusted Handbook method for sub-state estimates. This publication provides more detailed estimates for sub-state areas. In all, monthly labor force estimates are provided for 5,300 areas, including Metropolitan Statistical Areas (MSA's), Labor Market Areas (LMA's), all counties (cities and towns in New England), and cities of 25,000 population or more.

Estimates for all but the 11 largest states, Los Angeles, and New York City are revised a month following the initial publication in which they appear, and again at the end of the year. The first revision takes into consideration revisions to the Payroll Employment and Unemployment Insurance statistics. The end of year revision adjusts the monthly estimates such that their annual average equals the CPS sample based annual average estimates for those states and sub-state areas for which CPS data are provided.

5.2.1 Design of the Current Population Survey

The CPS monthly sample consists of 72,000 housing units. This sample size was chosen to meet national and state reliability requirements. Assuming a 6% unemployment rate, the national sample size was chosen so that a month-to-month change of 0.2 percentage points in the unemployment rate would be statistically significant at the 90 percent confidence level. This translates to a coefficient of variation (CV) of 1.8% for the national unemployment rate. The 11 largest states have a CV of 8.0% on the monthly unemployment rate. The other 39 states and the District of Columbia have a CV of 8.0% on the annual average unemployment rate.

The CPS sample is located in 729 areas comprising over 1,000 counties and independent cities with coverage in every state and the District of Columbia. Prior to 1984, the CPS had been designed as a national sample with the goal of providing the best estimates of employment and unemployment for the U.S. as a whole.

The CPS sample is selected by first dividing the entire area of the United States into 1,973 primary sampling units (PSU's), where a PSU is a county or a number of contiguous counties. The 1,973 PSU's are grouped into strata within each state. One PSU is selected from each stratum with probability of selection proportionate to the population size of the PSU. The most populated PSU's are grouped by themselves and selected with certainty. Since the sample design is state based, the sampling ratio differs by state, ranging roughly from 1 in every 200 households to 1 in every 2500 households. There are several stages of selecting the household units within PSU's. First, enumeration districts, which are administrative units and contain about 300 housing units, are ordered so that the sample would reflect the demographic and residential characteristics of the PSU. Within each enumeration district the housing units are sorted geographically and are grouped into clusters of approximately four housing units. A systematic sample of these clusters of housing units is then selected.

Part of the sample is changed each month. For each sample, eight systematic subsamples (rotation groups) or segments are identified. A given rotation group is interviewed for a total of 8 months -- 4 consecutive months in the survey, followed by 8 months out of the survey, followed by 4 more consecutive months in the survey. Under this system, 75 percent of the sample segments are common from month-to-month and 50 percent of the sample segments are common from year-to-year.

The estimation procedures involves weighting the data from each sample person by the inverse of the probability of the person being in the sample. These estimates are then adjusted for noninterviews, followed by two ratio estimation procedures to adjust the CPS estimates to known population totals. The last step in the preparation of estimates makes use of a composite estimating procedure. The composite estimate for the CPS is a weighted average of the estimate for the current month and the estimate for the previous month, adjusted for the net month-to-month change in households.

Balanced repeated replication and collapsed stratum methods are used to estimate CPS variances for selected characteristics. Generalized variance functions are used to present the sampling error estimates in publications. Sampling error estimates are provided for all direct sample based estimates, which include the annual average estimates for states and some sub-state areas, as well as monthly estimates for the 11 largest states. Error estimates are not provided for estimates which use the State Model methodology or the Handbook methodology. General variance functions are used for calculating sampling error estimates for the direct sample based estimates from the CPS. The Employment and Earnings series provides methods for calculating sampling error estimates for almost any estimate in the publication. These methods can also be used to calculate estimates for unpublished CPS estimates, such as the monthly unemployment rates for the smaller states. These sampling error estimates can be used to approximate the error in the model based estimates.

5.3 Estimator Documentation

The method used to provide monthly state estimates for the 39 states and the District of Columbia is based on the time series approach to sample survey data. Originally suggested by Scott and Smith (1974), this approach treats the population values as stochastic and uses signal extraction techniques developed in the time series literature to improve on the direct survey estimator. Recent work has been conducted by Bell and Hillmer (1990), Binder and Dick (1990), Pfefferman (1992), and Tiller (1992a).

The actual monthly CPS sample estimates are represented in signal plus noise form as the sum of a stochastically varying true labor force series (signal) and error (noise) generated by sampling only a portion of the total population. Issues related to non-sampling errors are not considered by this approach. The signal is represented by a time series model that incorporates historical relationships in the monthly CPS estimates along with auxiliary data from the Unemployment Insurance (UI) and Current Employment Statistics (CES) programs. This time series model is combined with a noise model that reflects key characteristics of the sampling error to produce estimates of the true labor force values. This estimator has been shown to be design consistent under general conditions by Bell and Hillmer (1990) and is optimal under the model assumptions.

Unlike the typical small area estimation application that seeks to improve on the direct survey estimator by borrowing strength over areas, the time series approach borrows strength over time for a given area. While variance reduction is a primary goal of both these approaches, when there are strong overlaps in the sample design and relatively long historical series are available, the time series approach provides powerful tools for estimating the underlying population values. As discussed in the previous section, the CPS design creates major sample overlaps resulting in very strong autocorrelations in the sampling errors. By combining a model of both the true labor force values and the sampling error, the time series approach controls for the autocorrelation induced by the sample design making it easier to identify the population dynamics. This is particularly useful in trend analysis and seasonal adjustment. When sampling error is strongly autocorrelated, trend and sampling effects are confounded in the observed data (Tiller, 1992b).

A structural time series model with explanatory variables (Harvey) is used to model the true values of the employment level and the unemployment rate for the 39 states and the District of Columbia. The observed CPS labor force estimate, Y_t, is represented as the sum of the signal, θ_t, plus a noise term, e_t.

$$Y(t) = \theta(t) + e(t) \tag{5.1}$$

The signal is modeled as a time series decomposed into the form

$$\theta(t) = X(t)\beta(t) + T(t) + S(t) + I(t) \tag{5.2}$$

where the terms on the right-hand side denote the regressors with time varying coefficients, trend, seasonal, and irregular components of the signal at time t. The first three components are allowed to drift slowly over time by subjecting them to mutually independent, normally distributed white noise disturbances. The variances of these disturbances constitute the hyperparameters of the signal and determine the stochastic properties of the individual components. A positive variance for a component implies that it is a stochastic process, possibly nonstationary, while a zero variance implies deterministic behavior. The irregular is treated as stationary. These components are explained below.

Regressor component
A dynamic linear regression model is used to represent that part of the signal explained by a set of observable economic variables

$$X(t)\beta(t) \tag{5.3}$$

where $X(t)$ is a 1 x k vector of the known explanatory variables and $\beta(t)$ is a k x 1 coefficient vector modeled as a random walk where $v_\beta(t)$ is a multivariate normal vector of mutually independent, zero mean random shifts

$$\beta(t) = \beta(t-1) + v_\beta(t)$$

$$\tag{5.4}$$

$$E\left[v_\beta(t)v_\beta(t)\right] = \text{Diag}\left(\sigma^2_{\beta(1)}, \ldots, \sigma^2_{\beta(k)}\right).$$

The presence of these variables serve two important and related functions. First, it allows the use of auxiliary data obtained through administrative and other non-CPS sources to improve the efficiency of model estimates. Secondly, as economic indicators these variables play a useful descriptive function that helps the state analysts explain their labor force movements.

A common core of regressor (explanatory) variables have been developed for the District of Columbia and the 39 state unemployment rate models. Each state model is based on two non-CPS state specific data sources -- UI claims and CES nonagricultural payroll employment data. To control for important cyclical and seasonal labor force movements not accounted for by the UI and CES data, variables have been constructed from selected CPS data in such a way as to reduce the influence of sampling error. The CPS regressor variables include the employment to population ratio and the entrant (into the labor force) unemployment rate. A similar set of regressor variables have been developed for the employment models.

The stability of the regression coefficients, $\beta(t)$, depend upon their respective variances, σ^2_β. A zero variance results in a fixed coefficient while a positive variance allows the coefficient to change smoothly over time.

The role of the stochastic time series components, T_t, S_t, and I_t is to represent systematic variation in the signal not accounted for by the auxiliary variables.

Trend component
Low frequency behavior is represented by a local approximation to a linear trend

$$T(t) = T(t-1) + R(t-1) + v_T(t)$$

$$(5.5)$$

$$R(t) = R(t-1) + v_R(t)$$

where $v_T(t)$ and $v_R(t)$ are mutually independent white noise disturbances with variance σ^2_{vT} and σ^2_{vR}, respectively. $T(t)$ and $R(t)$ can be interpreted as the level and slope variables associated with the locally smooth trend. If both variances are zero, a fixed linear trend results.

Seasonal component
The seasonal component is the sum of six trigonometric terms associated with the 12-month frequency and its five harmonics

$$S(t) = \sum_{j=1}^{6} S_j(t) \tag{5.6}$$

where each of the individual terms $S_j(t)$ is subject to a white noise shock, $v_{s_j}(t)$, assumed to have a common variance, σ_s^2.

$$S_j(t) = \cos(w_j) S_j(t-1) + \sin(w_j) S_j^*(t-1) + v_{s_j}(t)$$

$$\tag{5.7}$$

$$S_j^*(t) = \sin(\omega_j) S_j(t-1) + \cos(\omega_j) S_j^*(t-1) + v_{s_j}^*(t)$$

$$\omega_j = 2\pi \rho_j^{-1}, \rho = 12, \ 6, \ 4, \ 3, \ 2.4, \ 2$$

Over a 12-month period the expected seasonal effects add to zero

$$E\left[\sum_{\lambda=0}^{11} S_{t-\lambda} \right] = 0. \tag{5.8}$$

A positive value for σ_s^2 permits the seasonal pattern to evolve over time while a zero value results in a fixed seasonal pattern.

Irregular component
The irregular component is a residual not explained by the regression or time series components discussed above. The convention in classical decomposition of a univariate time series is to represent the irregular as a highly transient phenomena, i.e., as white noise or a low order MA process.

Noise
The noise component of the observed CPS estimate represents error that arises from sampling only a portion of the total population. Its structure depends upon the CPS design and population characteristics. For our purposes, we focus on those design features that are likely to have a major effect on the variance-covariance structure of the sampling error, $e(t)$.

One of the most important features of the CPS is the large overlap in sample units from month to month. As described in the previous section, units are partially replaced each month according to a 4-8-4 rotating panel. Since this system produces large overlaps between samples one month and one year apart, we can expect $e(t)$ to be strongly autocorrelated. Also, there is likely to be some correlation between nonidentical units in the same rotation group because of the way in which new samples are generated. When a cluster of housing units permanently drops out of a rotation group, it is replaced by nearby units. Since the new units will have characteristics similar to those being replaced, this will result in correlations between nonidentical households in the same rotation group (Train, Cahoon, and Makens, 1978).

Finally, the dynamics of the sample error will also be affected by the composite estimator. This is a weighted average of an estimate based on the entire sample for the current month only and an estimate which is a sum of the prior month composite and change that occurred in the six rotation groups common to both months (Bureau of the Census, 1978). In effect, this estimator takes a weighted average of sample data from the current and all previous months.

Another important feature of the CPS is its changing variance over time. There are three major sources of heteroscedasticity: (1) sample redesigns; (2) changes in the sample size; and (3) changes in the true value of the population characteristic of interest. The first two cause discrete shifts in the sample variance. For example, the CPS is redesigned each decade to make use of decennial census data to update the sampling frame and estimation procedures. Most recently, a state-based design was phased in during 1984/85 along with improved procedures for noninterviews ratio adjustments and compositing. Changes in state sample sizes have occurred more frequently than redesigns and have had major effect on variances at the state level. Even with a fixed design and sample size, the error variance will be changing because it is a function of the size of the true labor force. Since the labor force is both highly cyclical and seasonal, we can expect the variance to follow a similar pattern. To capture the autocorrelated and heteroscedastic structure of $e(t)$, we may express it in multiplicative form (see Bell and Hillmer, 1990) as

$$e(t) = \gamma(t) e^*(t) \qquad (5.9)$$

with $e^*(t)$ reflecting the autocovariance structure, assumed to follow an ARMA process and $\gamma(t)$ representing a changing variance over time. More explicity

$$e^*(t) = \phi_e^{-1}(L)\theta_e(L)v_e(t)$$

$$\gamma(t) = \sigma_e(t)/\sigma_e.$$

where

$\sigma^2_{e_*}$ = the sampling variance at time t

$\theta_e(L)$ = a stationary moving-average operator of order q_e

$\phi_e(L)$ = a stationary autoregressive operator of order p_e

$v_e(t)$ = a white noise disturbance

L is the lag operator such that $L^k\left(y(t)\right) = y(t-k)$

$$\sigma^2_{e_*} = \sigma^2_{v_e} \sum_{k=1}^{\infty} g_k$$

The weights $\{g_k\}$ are computed from the generating function,

$$g(L) = \phi_e^{-1}(L)\theta_e(L).$$

The autocovariance structure may also change over time with redesigns of the sample. However, since the most important source of autocorrelation is the 4-8-4 rotation scheme, which has not changed, it seems reasonable to treat this structure as stable, at least, between sample designs.

The application of the signal-plus-noise approach requires information on the variance-covariance structure of the CPS at the state level. In principal, this information can be estimated directly from the sample unit data using conventional designed based methods. In practice, this is not always feasible, since the CPS variance estimation involves complex computations on large microdata files. In the initial implementation of models in 1989 for the 39 states and the District of Columbia not enough information was available to explicitly model the sampling error. Instead, the noise component was estimated as a correlated residual (Tiller, 1989). More recently, sampling error autocorrelations have been developed and new models are being tested incorporating this information (Tiller, 1992a).

Estimation

The models described in the previous section are estimated using the Kalman filter (KF). The KF is a highly efficient algorithm for estimating unobserved components of a time series model, when that model can be represented in state-space form. The state-space form consists of two sets of equations, transition and measurement equations, and a set of initial conditions. The unobserved signal and noise components are collected into the state vector, $Z(t)$. The transition equations represent the state vector as a first-order vector-autoregressive process (VAR) with a normal and independently distributed disturbance vector, $v(t)$, which contains the white noise disturbances associated with each of the unobserved component processes. The transition equations are set out below in a simplified form appropriate for our specific application.

$$\underset{mx1}{Z(t)} = \underset{mxm}{F} \ \underset{}{Z(t-1)} + \underset{mxj}{G} \ \underset{jx1}{v(t)} \qquad (5.10)$$

$$v(t) \sim NID(0, \Omega), \underset{jxj}{\Omega} = DIAG\left(\sigma_{v_j}^2\right)$$

$$Z(0) \sim N[Z0, P0]$$

where F and G are nonstochastic matrices, and $Z(0)$ is the initial state vector, assumed to be distributed independently of $v(t)$. The regression coefficients, trend, seasonal, and irregular components easily fit into the state-space form since they are already expressed as first order AR processes. The ARMA form of the sampling error is easily translated into a VAR form and, then, included in the state vector along with the other components.

The measurement equation represents the observed sample value at time t as a linear combination of the unobserved state variables

$$y(t) = \underset{1xm}{H(t)Z(t)} \qquad (5.11)$$

where $H(t)$ is the nonstochastic observation vector containing the explanatory variables, variance inflation factors, $\gamma(t)$, and zero or one values to select other components that directly sum to the sample estimate. The signal and noise components are retrieved from the state vector by applying the appropriate linear combination to the components of the state vector.

$$\theta(t) = H_\theta(t)Z(t)$$

$$e(t) = H_e(t)Z(t)$$

(5.12)

The vector, $H_\theta(t)$, is identical to H(t), except that the components of $Z(t)$ not associated with the signal are zeroed out. Similarly, $H_e(t)$ acts to select out the noise components. In a like manner, subcomponents of interest, such as the trend and seasonal, can be extracted from the state vector.

In the new models under development, the CPS error correlation structure is estimated outside of the time series model from design-based information. Variances for the state CPS estimates are computed using the method of generalized variances (Tiller, 1992). Autocorrelations were derived in a study by Dempster and Hwang (1992). In that study, state-specific variance component models were fit to a time series of data for the 8 CPS rotation groups. From the estimated variance parameters, autocorrelations were derived and, then, ARMA parameters were estimated from these autocorrelations.

The state-space system contains two sets of unknowns (the sampling error lag covariances are predetermined, as discussed above): the variance parameters of the signal components, Ω, and the unobserved state vector, $Z(t)$. The KF is used to estimate both sets of unknowns. First, consider the estimation of $Z(t)$ given Ω. Since $V(t)$ is independent multivariate normal, the conditional distribution of $Z(t)$ given current and/or past values of $y(t)$, is also normal,

$$Z\left[t \,/\, y(t-1)...y(1)\right] \sim N\left[Z(t\,/\,t-l),\, P(t\,/\,t-l)\right],\, l \geq 0 \qquad (5.13)$$

where $Z(t\,/\,t-l)$ and $P(t\,/\,t-l)$ refer to the mean of $Z(t)$ and its covariance matrix, conditional on sample information up to time $t-l$. This conditional density provides the link between the unobserved state variables and the observable sample data. Given the model assumptions and Ω, the conditional mean is the optimal (minimum variance unbiased) estimator of $Z(t)$.

The KF provides a recursive formula for calculating the conditional mean of the state vector at time t, $Z(t/t)$, and its covariance matrix by means of updating the estimator, $Z(t/t-1)$. It is constructed from two sets of equations derived from the state-space equations -prediction equations and update equations. The prediction equations compute the mean vector and covariance matrix for the conditional density based on sample data prior to time t, the variance parameters, Ω, and the initial state vector $Z(0)$.

$$Z(t/t-1) = FZ(t-1/t-1)$$

$$P(t/t-1) = FP(t-1/t-1)F' + G\Omega G'$$

(5.14)

and the mean $y(t/t\text{-}l)$, and variance $f(t/t\text{-}l)$ of the conditional density of the sample observation is given by

$$y(t/t-1) = H(t)Z(t/t-1)$$

$$f(t/t-1) = H(t)P(t/t-1)H'(t)$$

(5.15)

Once an additional observation, $y(t)$, becomes available, the update equations revise the conditional moments with the new information in that observation.

$$Z(t/t) = Z(t/t-1) + k(t)\upsilon(t/t-1)$$

$$P(t/t) = [I - k(t)H(t)]P(t/t-1)$$

(5.16)

where

$$k(t) = [f(t/t-1)]^{-1}P(t/t-1)H'(t)$$

$$\upsilon(t/t-1) = y(t) - H(t)Z(t/t-1)$$

The quantity, $k(t)$, is the gain of the KF, $\upsilon(t/t\text{-}l)$ is the innovation of the filter at time t, which is the one-step-ahead error in predicting $y(t)$ with its conditional mean, $y(t/t\text{-}l)$.

The estimator of the signal and its covariance matrix are given by

$$\theta(t/t) = H_\theta(t)Z(t/t)$$

$$P_\theta(t/t) = H_\theta(t)P(t/t)H'_\theta(t).$$

(5.17)

Similar expressions exist for the noise component.

To initialize these equations, it is necessary to specify starting values for the conditional moments, $Z(0)$ and $P(0)$. Those elements of the state vector that are stationary, i.e., sampling error and the irregular, are initialized with their unconditional moments. The nonstationary and nonstochastic state variables are initialized with diffuse priors.

Together, the prediction and update equations constitute the KF. The KF updates its latest prediction of the state vector with current sample data, prepares a prediction for the next period and updates that prediction when new sample data become available, but the estimate of $Z(t)$ will not be revised with data later than period t. Thus, the KF estimator at time t is optimal only with respect to data later than period t. The estimator of $Z(t)$ optimal for all observations, before and after t, is known as a smoother. By taking a linear combination of a forward and backward KF, which is the KF run in reverse, starting at the end of the sample period at time $t=T$, and preceding to the beginning, a Kalman (fixed interval) smoother (KS) is obtained. Let the backward filter prediction of the state vector at time t, conditional on data from $t+1$ to T be denoted by $Z(t/t+1)$ and its covariance by $P(t/t+1)$. The smoothed estimator is

$$Z(t/T) = P(t/T)\left[P(t/t)^{-1} Z(t/t) + P(t/t+1)^{-1} Z(t/t+1) \right]$$

$$\text{(5.18)}$$

$$P(t/T) = \left[P(t/t)^{-1} + P(t/t+1)^{-1} \right]^{-1}$$

From the covariance expression for $Z(t/T)$, we have

$$P(t/T) = \left[P(t/t)^{-1} + P(t/t+1)^{-1} \right]^{-1},$$

which implies that $P(t/T) - P(t/t)$ is negative semidefinite. Thus, the smoothed estimator is at least as good as the KF estimator. The smoothed estimates of the signal and noise components are obtained from the appropriate linear combination of the conditional mean vector, $Z(t/T)$.

We now consider the estimation of Ω. Since the conditional density for a single observation, $y(t)$, is normal with mean, $y(t/t-l)$ and variance, $f(t/t-l)$, given directly by the KF, the joint density of the T sample observations is the product of these individual densities. If the state vector contains l nonstationary elements, then the log likelihood, conditional on the first l observations required to form priors for these nonstationary elements, may be expressed as

$$L\left[\Omega / y(l+1),...,y(T); Z0, P0\right] = \sum_{t=l+1}^{T} \ln f(t/t-1) + \left(y(t) - y(t/t-1)\right)^2 f(t/t-1)^{-1}$$

(5.19)

The maximum likelihood estimators of the variance components are obtained by maximizing L with respect to Ω. The KF is used to evaluate the likelihood for a given Ω and L is maximized by a nonlinear search of the parameter space performed using a quasi-Newton routine.

State agency staff prepare their official monthly estimates using software developed by BLS that implements the KF. This algorithm is particularly well suited for the preparation of current estimates as they become available each month. Since it is a recursive data processing algorithm, it does not require all previous data to be kept in storage and reprocessed every time a new sample observation becomes available. All that is required is an estimate of the state vector and its covariance matrix for the previous month. The software is interactive, querying users for their UI and CPS data and, then combining these data with CPS estimates to produce model based estimates. At the end of the year, the monthly KF estimates are revised, along with previous year estimates with the smoothing algorithm.

5.4 Evaluation Practices

At BLS, various time series regression methods were developed in the late 1970's and early 1980's in the hopes of replacing the Handbook methods. However, these methods were never implemented for a variety of operational and technical issues. The state space model work, which began around 1987, provided for considerable flexibility in specifying the signal component. It includes as special cases two classes of model based approaches to sample surveys that have appeared in the literature: 1) If the variances of regression coefficients are set to zero, e_t is uncorrelated, and I_t is white noise, the system reduces to Eriksen's sample regression model. In this case, the signal extraction problem is solved by fitting a weighted least squares equation to observed sample data.   2) By specifying I_t as an ARIMA process and dropping the structural components, we have a class of models based on signal extraction theory. The regression mean drops out and the signal reduces to a covariance stationary differenced process. If in addition, the variance and the ARMA parameters of the e_t process are also held constant, then e_t will also be covariance stationary. Hence the development of best models within the state space formulation implicitly includes comparisons against the time series methods developed earlier.

For each of the 39 states and the District of Columbia, signal plus noise models of the CPS unemployment rate and employment level were fit to monthly data beginning in 1976. Each of the 80 models has been subjected to a wide variety of statistical tests.

An analysis of the model's prediction errors is the primary tool for assessing model adequacy. The prediction errors are computed as the difference between the current values of the CPS and the predictions of the CPS made from the model based on data prior to the current period. Since these errors represent movements not explained by the model, they should not contain any systematic information about the behavior of the signal or noise component of the CPS. Specifically, the prediction errors, when standardized, should approximate a randomly-distributed variate with zero mean and constant variance (white noise). The tests used to check the prediction errors for departure from these properties included:

- General tests for non-zero correlations in the innovations
- Departures from white noise behavior at the seasonal frequencies
- Heteroscedasticity
- Non-normality
- Prediction bias

About 50 to 60 percent of the total variance in the monthly CPS series is attributable to the estimated signal with the remainder due to the aggregate noise term. The time varying regression mean is considerably smoother than the underlying CPS series. Based on the diagnostic tests, the 80 models appear to fit the systematic underlying movements in the CPS fairly well. The major problems with the models were high autocorrelations in 11 states, and heteroscedasticity in 9 of the 40 unemployment rate models. The heteroscedasticity is in part a reflection of changing variances (and sample sizes) in the CPS. Explicitly modeling the CPS sample errors would alleviate this problem and is discussed in the current problems and activities section, below.

The current state model estimates were introduced in January 1989. The previous Handbook method could be classified as an accounting method. Several years of research and development, beginning in the early 1980's, examined numerous regression and time series approaches to replace the accounting method. A number of workgroups were setup to determine the criteria to be used to select the new methodology as well as how to implement the new methodology.

Ongoing evaluation of models includes annual reassessment of the regressor variables, if requested by staff in the state employment security agencies. Typically, state agency staff express concerns with the models if either the month to month movements in the unemployment rate estimates are larger than they expect or the unemployment rate level seems unreasonable compared to other economic data. Diagnostics tests, similar to the ones used for developing the model, are run. Adjustments to the regressor variables may be made if the diagnostic indicate a problem with the model. In this case historical estimates would be replaced, in addition to developing a new model for concurrent estimates.

5.5 Current Problems and Activities

The implementation of model estimates for states in January 1989 resulted, not unexpectedly, in estimates with more month to month volatility than the previous Handbook method. The previous method incorporated a 6 month moving average, which limited month to month movement. The seasonal variation in the employment and unemployment statistics series is usually large relative to the trend and cycle. However, the BLS decided to conduct a research project to investigate the issues of seasonally adjusting model based estimates prior to implementing seasonally adjusted state estimates. In November 1989, a work group was formed to examine issues related to the seasonal adjustment of estimates of employment level and unemployment rate for the 39 smaller states and the District of Columbia (non-direct use states). The group was charged with addressing two primary areas:

1. Evaluation of the performance of the model estimates relative to the CPS sample-based estimates, with emphasis on the trend/cycle characteristics of the series.

2. Evaluation of the use and limitations of the BLS standard seasonal adjustment method, X-11 ARIMA, to seasonally adjust the model estimates.

The evaluation of the modeling approach involved simulating a reduction in reliability of direct-use CPS samples in 2 large states (direct use -- Florida and Massachusetts) to nondirect-use levels, fitting models to the resulting weakened series, and then comparing model estimates to the CPS estimates from the full sample. While it would have been desirable to simulate sample cuts by subsampling the original data, this was considered too costly. Instead random noise was added to the full sample estimates, using an estimated variance/covariance structure of the CPS estimator. For each state, two weakened samples were generated for employment and unemployment, and separate models were fitted to the full and weakened samples. The main findings are summarized as follows:

Model Evaluation

1. Modeling the weakened unemployment rate series resulted in estimates which were closer to the full CPS sample than the unmodeled weak CPS series for all four unemployment rate series. Values for the root mean squared relative difference (RMSRD) comparing full CPS to model estimates were 28 to 38 percent smaller than values of RMSRD comparing the full CPS to the weakened series. Modeling also reduced the number of weakened series estimates falling outside two standard deviation intervals about the full sample estimates by 50 to 75 percent.

2. Modeling the weakened employment series resulted in modest, if any, reductions in the RMSRD from the full CPS sample. In one case for the Florida employment series, the RMSRD values for the model were actually larger than those for the weakened series. This appeared to be due primarily to the fact that the difference in reliability between the full sample and the weakened series for employment was very small compared to the difference for unemployment rate.

3. Modeling dramatically reduced the magnitude of irregular fluctuation in both employment and unemployment rate. It was not unusual for the relative contribution of the variance of monthly change in the CPS to be 4 to 8 times that of the modeled series. The much smoother quality of the model estimates have important implications for seasonal adjustment (see below).

Evaluation of Seasonal Adjustment

Using X-11 ARIMA, the CPS and modeled series were seasonally adjusted. The adequacy of the seasonal adjustment was evaluated using X-11 ARIMA quality control statistics, spectral analysis, sliding spans, and graphs of seasonal factors. The major findings are as follows:

1. The direct sample based unemployment rate CPS series could not be adequately seasonally adjusted. Frequently, several of the X-11 ARIMA quality control statistics failed. Seasonal adjustments had poor stability properties, the seasonal variation could not be completely removed, and distortion was added to the nonseasonal variation in the series. The results were better for the CPS employment series but not nearly as good as for the modeled series.

2. The seasonal adjustments for all four of the employment and unemployment model series appeared satisfactory. Spectral analysis shows that X-11 ARIMA was able to effectively remove seasonal variation in the modeled series without introducing distortions in the nonseasonal components of the series. The sliding span statistics indicate seasonal factors remain stable as the span of the data is shifted across time. In addition, monthly seasonal factors using the model were similar to the seasonal factors of the full sample estimates of the two direct-use states. This indicates that the models were not forcing an artificial pattern, but were "picking up" the seasonal pattern of the underlying CPS series, despite the extra noise which was introduced.

In summary, despite some limitations to the methods of evaluations, the study provided important information to help understand the value of modeling and the use of X-11 ARIMA to seasonally adjust model-based estimates; however, the theoretical base for superimposing the modeling structure for X-11 ARIMA to already smoothed, model based values remains to be explored. Although the study confirmed support for modeling, further work will be done to demonstrate the utility of the employment models.

The BLS introduced seasonally adjusted state employment and unemployment estimates beginning in January 1992, based on the results of this study.

Current research is focusing on further reduction in irregular movement in employment and unemployment models by introducing several changes to the methods. The most important change is the inclusion of the variance/covariance structure of CPS estimates into the models, rather than relying on the model to make these estimates (Tiller, 1992a). Information on the structure of the CPS sample error is being used to decompose the disturbance term into its sample error and model error components. Given CPS error variances and lag covariance, ARMA models can be developed to approximate the time series behavior of the sampling error. Treating the ARMA coefficients as known parameters of the state space system, standard time series diagnostic tools may be used to model the errors in equation disturbances. The need for estimating the variance-covariance structure of the CPS estimates stems from sample redesign and changes in sample size.

Other changes, such as removing some exogenous CPS variables, are expected to improve the seasonal movements in the model estimates. Florida and Massachusetts will again be used to examine the ability of the weakened series to track the full sample CPS estimates. Research is expected to be completed in FY93 for implementation in January 1994.

Long term research will focus on substate estimation. Hierarchical and Empirical Bayes methods may be considered in addition to a time-series approach for substate estimates. Spatial models, which borrow strength from CPS sample data within the state, may be appropriate for substate estimates.

A number of related studies have been conducted under the auspices of other governmental agencies. The Census Bureau has supported research by Bell and Hillmer (1990) that has been instrumental in stimulating renewed interest in the time series approach to survey estimation. In this study, the authors applied ARIMA models to retail survey data. Binder and Dick (1990), at Statistics Canada, fitted ARMA models to Canadian Labour Force survey data. Both of these studies estimated the sampling error structure outside the time series model, using design-based methods.

Pffermann (1991) applied a structural time series model to individual panel estimates from the Israeli labor force survey. The sampling error structure was estimated through the model rather than by design-based methods.

Dempster and Hwang (1993) have developed prototype Bayesian models for estimating U.S. State employment and unemployment rates. Their basic time series models are constructed from fractional Gaussian noise processes.

REFERENCES

Bell, W.R. and Hillmer, S.C. (1990), "The Time Series Approach to Estimation for Repeated Surveys". Survey Methodology, 16, 195-215.

Binder, D.A. and Dick, J.P. (1990), "A Method for the Analysis of Seasonal ARIMA Models," Survey Methodology, 16, 239-253.

Bureau of Labor Statistics (1988), Handbook of Methods, Washington, D.C.

Bureau of Labor Statistics (1991), Report on the Seasonal Adjustment of LAUS Model Estimates, Washington, D.C.

Bureau of Labor Statistics (1991), The Current Population Survey - An Overview, Internal Document by Edwin Robison, Washington, D.C.

Bureau of the Census (1978), The Current Population Survey: Design and Methodology, Technical Paper 40, Washington, D.C.

Dempster, A.P. and Jing-Shiang Hwang (1993), "Component Models and Bayesian Technology for Estimation of State Employment and Unemployment Rates," paper presented at the 1993 Annual Research Conference, Census Bureau.

Harvey, A.C. (1989), Forecasting Structural Time Series Models and the Kalman Filter, Cambridge University Press

Pfeffermann, D. (1992). Estimation and Seasonal Adjustment of Population Mean Using Data from Repeated Surveys. Journal of Business and Economics Statistics, 9, 163-175.

Scott, A.J. and Smith, T.M.F. (1974), "Analysis of Repeated Surveys Using Time Series Methods," Journal of the American Statistical Association, 69, 674-678.

Tiller, R. (1989), "A Kalman Filter Approach to Labor Force Estimation Using Survey Data," in Proceedings of the Survey Research Methods Section, American Statistical Association.

_____ (1992a), "Time Series Modeling of Sample Survey Data from the U.S. Current Population Survey," Journal of Official Statistics, 8, 149-166.

_____ (1992b), "A Time Series Approach to Small Area Estimation," in Proceedings of the Survey Methods Research Section, American Statistical Association.

Train, G., Cahoon, L., and Makens, P. (1978). The Current Population Survey Variances, Inter-Relationships, and Design Effects. In Proceedings of the Survey Research Methods Section, American Statistical Association, 443-448.

CHAPTER 6

County Estimation of Crop Acreage Using Satellite Data

Michael Bellow, Mitchell Graham, and William C. Iwig
National Agricultural Statistics Service

6.1 Introduction and Program History

The National Agricultural Statistics Service (NASS) of the U.S. Department of Agriculture (USDA) has published county estimates of crop acreage, crop production, crop yield and livestock inventories since 1917. These estimates assist the agricultural community in local agricultural decision making. Also the Federal Crop Insurance Corporation (FCIC) and the Agricultural Stabilization and Conservation Service (ASCS) of the USDA use NASS county crop yield estimates to administer their programs involving payments to farmers if crop yields are below certain levels. The primary source of data for these estimates has always been a large non-probability survey of U.S. farmers, ranchers, and agribusinesses who voluntarily provide information on a confidential basis (see Chapter 7). In addition, the U.S. Census of Agriculture, conducted by the Bureau of the Census every five years, serves as a valuable benchmark for the NASS county estimates.

Earth resources satellite data, particularly from the Landsat series of satellites, provide another useful ancillary data source for county estimates of crop acreage. The potential for improved estimation accuracy using satellite data is based on the fact that, with adequate coverage, all of the area within a county can be classified to a crop or ground cover type. The accuracy of the estimates is then dependent on how accurately the satellite data are classified to each crop type based on the "ground truth" data obtained from the annual June Agricultural Survey (JAS) conducted by NASS. Through the use of aerial photographs, this survey identifies the crop type of individual fields within randomly selected land segments. Segments in major agricultural areas are approximately one square mile in area and normally contain 10 to 20 fields. The satellite spectral data are matched to the corresponding fields for use in classifying all individual imaged areas, known as pixels, to a particular crop type. Recent studies (Bellow 1991; Bellow and Graham 1992) have shown that, for certain crops, more than 80 percent of the pixels are classified correctly. This correct classification level is high enough to provide improved estimation accuracy.

NASS has been a user of remote sensing products since the 1950's when it began using mid-altitude aerial photography to construct area sampling frames (ASF's) for the 48 states of the continental United States. A new era in remote sensing began in 1972 with the launch of the Landsat I earth-resource monitoring satellite. Four additional Landsats have been launched since 1972, with Landsat IV and V still in operation in 1993. The polar-orbiting Landsat satellites contain a multi-spectral scanner (MSS) that measures reflected energy in four bands of the electromagnetic spectrum for an area of just under one acre. The spectral bands were selected to be responsive to vegetation characteristics. In addition to the MSS sensor, Landsats IV and V have a Thematic Mapper (TM) sensor which measures seven energy bands and has increased spatial resolution. The large area (185 by 170 km) and repeat (16 day per satellite) coverage of these satellites opened new areas of remote sensing research: large area crop inventories, crop yields, land cover mapping, area frame stratification, and small area crop cover estimation.

Research from 1972 to 1978 led to the creation of an operational procedure for large area crop acreage estimation. A regression estimator was developed which related the ground-gathered area frame data to the computer classification of Landsat MSS images. The basic regression approach used to produce State estimates does not produce reliable county estimates. Domain indirect regression estimators were developed for this purpose. In the 1978 crop season, corn and soybean acreage State and county estimates based on remotely sensed data were produced for Iowa. One to two States were added to the project through 1984. For the 1984-1987 crop seasons, this project covered an eight-State area in the central United States and produced regression estimates of corn, winter wheat, soybeans, rice, and cotton acreages. These regression estimates were combined with other survey indications and administrative data to provide final published county estimates. Estimation based on data from Landsat MSS sensors was discontinued in 1988 in order to implement the increased capabilities of higher resolution sensors.

France entered the field of earth resources satellites in 1986 with the launch of SPOT I, which carries an improved multi-spectral scanner. This scanner images an even smaller area than the TM sensor but only measures three energy bands. Several NASS research projects compared the SPOT MSS and Landsat TM sensors with respect to crop estimation. This research led to the selection of Landsat TM as the preferred sensor for crop area estimation based on its superior spectral characteristics. The spatial characteristics of the SPOT MSS sensor provide a benefit only in areas with mostly small fields.

Regression estimation of crop acreages for large and small areas based on computer classification was reinstated in 1991 with the Delta Remote Sensing Project using Landsat Thematic Mapper data imaged over the Mississippi Delta region, which is a major rice and cotton area. Results from the operational eight-State program in 1987 and from sensor comparison experiments showed that the regression approach was most effective for rice and cotton estimation. State and county estimates of rice, cotton, and soybean acreages were produced for Arkansas and Mississippi in 1991, with Louisiana added in 1992. The project only covered Arkansas in 1993 due to budgetary constraints.

Three domain indirect regression estimators have been used or considered for producing small area county estimates using ancillary satellite data. From 1976 to 1982, the Huddleston-Ray estimator was used (Appendix B). In 1978, the Cardenas family of estimators was considered but not implemented (Appendix C). Beginning in 1982, the Battese-Fuller family of estimators was used for calculating county crop acreage estimates using Landsat MSS data. Since 1991, the Battese-Fuller model has been used to produce county estimates with Landsat TM data. Currently, this is the preferred model. However, non-regression estimation procedures based on total pixel counts are being evaluated.

6.2 Program Description, Policies, and Practices

The basic element of Landsat spectral data is the set of measurements taken by a sensor of a square area on the earth's surface. The sensor measures the amount of radiant energy reflected from the surface in several bands of the electromagnetic spectrum. The individual imaged areas, known as pixels, are arrayed along east-west rows within the 185 kilometer wide north-to-south pass (swath) of the satellite. For purposes of easy data storage, the data within a swath are subdivided into overlapping square blocks, called scenes. The latest two satellites (Landsats IV and V) image a given point on the earth's surface once every 16 days. The MSS sensor, formerly used for crop area estimation, contained four spectral bands with 80 meter spatial resolution. The more advanced TM sensor has seven bands (three visible and four infrared) with 30 meter resolution.

Several Landsat scenes may be required to cover an entire region of interest within a given State. It is not always possible to have the same image date for all such scenes due to schedule, cloud cover, and image quality factors. Consequently, analysis districts are created. An analysis district is a collection of counties or parts of counties contained in one or more Landsat scenes that have the same image date, or in areas for which usable Landsat data is not available to the analyst. To obtain State level crop acreage estimates, NASS sums all analysis district level estimates within the State. County level estimates are obtained using domain indirect regression and synthetic estimation methods, to be discussed later.

The area sampling frame for each State is stratified based on land use such as percentage cultivation, forest, and rangeland. NASS uses the regression estimator described by Cochran (1977, pp. 189-204) to compute crop acreage estimates for each land use stratum within an analysis district that has satellite coverage for an adequate number of JAS segments. These regression estimates are more precise than the direct expansion estimates obtained from JAS data alone. A detailed description of the procedure involved is provided by Allen (1990). Briefly, the steps required are as follows:

1. A graphics oriented registration process associates Landsat pixels with JAS sampled segments.

2. JAS data for sampled segments are used to label each pixel within the segments to a crop or other cover type.

3. Labelled pixels are clustered based on their Landsat data values to develop discriminant functions (signatures) for each cover type.

4. The discriminant functions are used to classify each pixel within the sampled segments to a cover type.

5. The segment level classification results are used to develop regression relationships for each crop between the ground and satellite data within each land use stratum. For each stratum, the independent (regressor) variable is the number of pixels classified to that crop per segment, and the dependent variable is the JAS segment reported crop acreage.

6. All pixels within the analysis district are classified, using the discriminant functions developed in Step 3.

7. For each stratum, the mean number of pixels per segment classified for a given crop over all segments in the population is substituted into the corresponding regression equation to obtain the stratum level mean crop acreage per segment. This mean is multiplied by the known total number of segments in the stratum to obtain the stratum level crop acreage estimate.

8. The stratum level estimates are summed to obtain the analysis district level crop acreage estimate for the portion of the analysis district covered by satellite data.

For land use strata lacking satellite coverage of an adequate number of JAS segments to develop the regression relationship, the direct expansion of JAS data is used to obtain estimates. These stratum level JAS estimates are also summed to obtain analysis district estimates for each crop representing the area not covered by satellite data. The total analysis district estimate for a particular crop is then:

$$\hat{T}_a = \hat{T}_{(REG),a.} + \hat{T}_{(DE),a.}$$

where:

$\hat{T}_{(REG),a.}$ = sum of the regression estimates over the land use strata with satellite coverage for analysis district a

$\hat{T}_{(DE),a.}$ = sum of the JAS direct expansion estimates over the land use strata without satellite coverage for analysis district a.

The final estimate of total crop acreage in the State is made by adding the analysis district estimates:

$$\hat{T}_s = \Sigma \, \hat{T}_a .$$

In many States, counties typically contain fewer than five sampled JAS segments, and may contain no segments at all. This fact makes it generally infeasible to define analysis districts to be individual counties and then use the above procedure to obtain county level estimates. Instead, the Huddleston-Ray, Cardenas, and Battese-Fuller domain indirect regression estimators have been developed and investigated for providing county estimates of crop acreage. The Battese-Fuller approach is currently favored by NASS, and is described in detail in Section 6.3.

The NASS County Estimates System, described in Chapter 7, is designed to accept the Battese-Fuller values as a separate set of county crop acreage estimates. Within this system, the Battese-Fuller county estimates are first scaled to be additive to the official NASS State estimate for each commodity. The scaled Battese-Fuller values are then composited with scaled values from other NASS surveys and administrative data sources. Thus the Battese-Fuller estimates serve as an additional input to the County Estimates System in States where they are available. Currently, the composite weights are subjectively set by the statisticians in the State office to provide satisfactory and reliable estimates. Each NASS State Statistical Office (SSO) prepares their own annual publication of the final county estimates. Although sampling variances are calculated for the Battese-Fuller estimates, no variances or error information are published for the final county estimates. Mean squared error information is only published for major agricultural items at the U.S. level.

6.3 Estimator Documentation

The Battese-Fuller family of estimators was first developed in the general framework of linear models with nested error structure (Fuller and Battese 1973), and later applied to the special case of county crop area estimation (Battese, Harter, and Fuller 1988). The method has been used for all Landsat county estimation done by NASS since 1982.

Similar to the State level estimation, land use strata are separated into those that have adequate satellite coverage and those that do not. The Battese-Fuller model can be applied within an analysis district for all strata where classification and regression have been performed. The analyst computes stratum level Battese-Fuller acreage estimates for all counties and subcounties within the boundaries of each analysis district. For land use strata where regression cannot be done due to lack of adequate satellite coverage or too few segments, a domain indirect synthetic estimator is used to obtain county estimates.

For a given analysis district, the strata where regression is performed will be referred to as regression strata and the remaining ones as synthetic strata. For convenience, the regression strata will be labelled $h=1,...,H_r$ and the synthetic strata $h=H_r+1,...,H$, where H_r is the number of regression strata and H the total number of strata in the analysis district. If a given county is only partially contained in the analysis district, the county estimation formulas given below apply only to the included portion.

6.3.1 Application of Battese-Fuller Model within Regression Strata

The Battese-Fuller model for county level estimation assumes that segments grouped by county admit the same slope relationship as the analysis district but that a different intercept is required. For each sample segment within regression stratum h in county c, the Battese-Fuller model proposes the following relation at the analysis district level:

$$
\begin{aligned}
y_{hci} &= \beta_{0h} + \beta_{1h} x_{hci} + \omega_{hci} \\
&= \beta_{0h} + \beta_{1h} x_{hci} + v_{hc} + \epsilon_{hci}, \quad i=1,...,n_{hc}
\end{aligned}
$$

where:

n_{hc} = number of sample segments in stratum h, county c

y_{hci} = JAS reported acres of crop of interest in stratum h, county c, segment i

x_{hci} = Landsat number of pixels classified to crop of interest in stratum h, county c, segment i

ω_{hci} =   regression error for stratum h, county c, segment i.

The two components of the error term ω_{hci} are the county effect v_{hc} and the

random error ϵ_{hci}.   They are assumed to be independent and normally distributed, with mean 0 and variances σ^2_{vh} and σ^2_{eh}, respectively. The covariance structure of the regression error terms is then:

$$\text{cov}(\ \omega_{hci} \ , \ \omega_{hkm} \ ) = 0, \qquad\qquad \text{if } c \neq k$$

$$= \sigma^2_{vh} \qquad\qquad \text{if } c=k, \ i \neq m$$

$$= \sigma^2_{vh}+\sigma^2_{eh}, \qquad\qquad \text{if } c=k, \ i=m$$

The parameter σ^2_{vh} is both a within county co-variance and a between county component of the variance of any residual, while σ^2_{eh} is the within county variance component for stratum h. The county mean residuals are observable and given by:

$$\bar{u}_{hc.} = \bar{y}_{hc.} - \hat{\beta}_{0h} - \hat{\beta}_{1h}\bar{x}_{hc.}$$

where:

$$\bar{y}_{hc.} = (1/n_{hc}) \sum_{i=1}^{n_{hc}} y_{hci}$$

and

$$\bar{x}_{hc.} = (1/n_{hc}) \sum_{i=1}^{n_{hc}} x_{hci}$$

and

$\hat{\beta}_{0h}, \hat{\beta}_{1h}$ =   least squares regression parameter estimates for stratum h.

For a given county the stratum level mean crop area per segment is estimated by:

$$\bar{y}_{(BF),hc.} = \hat{\beta}_{0h} + \hat{\beta}_{1h}\bar{x}_{hc.} + \delta_{hc}\bar{u}_{hc.}$$

where:

$\bar{x}_{hc}$ =   mean number of pixels per segment classified to crop in stratum h, county c (i.e., = X_{hc} / N_{hc})

X_{hc} =   number of pixels classified to crop in stratum h, county c

N_{hc} =   number of segments in stratum h, county c

δ_{hc} =   weighting factor selected by user ($0 \leq \delta_{hc} \leq 1$).

The total number of pixels classified to the crop of interest in stratum h, county c (X_{hc}) is provided by the analysis software. The total number of segments in stratum h, county c is obtained during the development of the area sampling frame.

The mean square error of this estimator is:

$$MSE\ (\bar{y}_{(BF),hc.}) = (1-\delta_{hc})^2 \sigma^2_{vh} + \delta_{hc}^2 (\sigma^2_{eh}/n_{hc})$$

This expression is conditional on known values of σ^2_{vh}, σ^2_{eh}, and δ_{hc}. The estimator of the mean square error is known to have a downward bias due to estimation of the variance components, and a correction due to Prasad and Rao (1990) is currently under investigation. In general, these parameters are unknown and will be estimated as discussed below.

Conditioned on the county effects, the mean squared bias is:

$$MSCB\ (\bar{y}_{(BF),hc.}) = (1-\delta_{hc})^2 \sigma_{vh}^2$$

The stratum level unadjusted estimator of total crop area in the county is:

$$\hat{T}_{(uBF),hc.} = N_{hc}[\beta_{0h} + \beta_{1h}\bar{x}_{hc} + \delta_{hc}\bar{u}_{hc.}]$$

The range of allowed values of δ_{hc} defines a family of Battese-Fuller estimators. If $\delta_{hc} = 0$, then the estimate of $\bar{y}_{(BF),hc.}$ lies on the analysis district regression line for the stratum. The mean square error for stratum h in county c is minimized by:

$$\delta_{hc}^* = n_{hc}\sigma^2_{vh}/(n_{hc}\sigma^2_{vh}+\sigma^2_{eh})$$

In general, the variance components σ^2_{vh} and σ^2_{eh} are unknown, so they must be estimated to approximate δ_{hc}^* using the above formula. The estimators currently used (Appendix A) represent a special case of the more general unbiased estimators derived by Fuller and Battese [1973], using the "fitting-of-constants" method. They require that a stratum contain at least two sample segments within the county in question. If there are fewer than two segments, then δ_{hc} is set to zero in the computation of the county estimate.

The unadjusted estimates of county totals generally do not sum to the corresponding analysis district totals obtained from large scale estimation. If county estimates are generated for an entire State, adjustment terms are used to ensure the county estimate sums match the corresponding district and State estimates. If county estimates are only provided for a region of the State, the unadjusted estimates are considered the best estimates. The resulting adjusted Battese-Fuller estimator is:

$$\hat{T}_{(aBF),hc.} = \hat{T}_{(uBF),hc.} - (N_{hc}/N_h) \sum_{j=1}^{C} \delta_{hj} \bar{u}_{hj.}$$

where:

N_h =   number of segments in stratum h.

The adjusted estimates sum to the appropriate analysis district totals. The estimate of total crop area in the regression strata of county c is:

$$\hat{T}_{(BF),c} = \sum_{h=1}^{H_r} \hat{T}_{(uBF),hc.} \quad (unadjusted\ form)$$

or

$$\hat{T}_{(BF),c} = \sum_{h=1}^{H_r} \hat{T}_{(aBF),hc.} \quad (adjusted\ form)$$

Estimation of the mean square error, bias, and variance of the unadjusted Battese-Fuller estimator can be accomplished via substitution of the selected values of δ_{hc} and estimates of σ^2_{vh} and σ^2_{eh} into the formulas for mean square error and mean square conditional bias given above (Walker and Sigman, 1982).

6.3.2 Application of Synthetic Estimator within Synthetic Strata

As mentioned earlier, domain indirect synthetic estimation is used to obtain crop area estimates for land use strata where regression is not viable. Since each county usually contains very few segments if any for a given stratum, the stratum level mean crop acreage per segment over the entire analysis district is used to compute the synthetic estimates. For synthetic stratum h, the estimate of crop area in county c is:

$$\hat{T}_{(SYN),hc.} = N_{hc}\bar{y}_{h..}$$

where:

$$\bar{y}_{h..} \quad = \quad \text{mean reported crop area per segment over all counties in stratum h.}$$

The domain indirect synthetic estimate of total crop area in the synthetic strata of county c is then:

$$\hat{T}_{(SYN),c} = \sum_{h=H_r+1}^{H} \hat{T}_{(SYN),hc.}$$

When doing the synthetic estimation, it is important to realize that the use of the analysis district level average to estimate county totals ignores county effects. Therefore, the synthetic component of a county estimate is often biased. In general, the larger and more heterogeneous the synthetic stratum, the greater will be this bias. It is therefore advantageous to use synthetic strata that have consistent agricultural intensity for the crops of interest.

6.3.3 Total County Estimate

The final county estimate is obtained by adding the regression and synthetic components together:

$$\hat{T}_c = \hat{T}_{(BF),c} + \hat{T}_{(SYN),c} .$$

If a county is split between two analysis districts, the estimates, $\hat{T}_{c,}$ from each district will be added to provide the total estimate.

6.4 County Estimates Example

This section describes an example of the computation of county estimates, using 1988 Iowa data. The example is taken from a research project that compared the TM sensor with the French SPOT multi-spectral scanner with regard to estimation effectiveness (Bellow 1991). However, county estimation was only done with the TM data. The research site was a nine county region in western Iowa, where corn and soybeans are the major crops. Ground data from the 1988 JAS were used, involving a sample of 30 segments. Of the 30 segments used for the study, 28 came from stratum A (agricultural) and the other two from stratum B (agri-urban).

Table 1: Population (N_{hc}) and Sample (n_{hc}) Number of Segments by County, Stratum

County	Stratum A		Stratum B	
	N_{hc}	n_{hc}	N_{hc}	n_{hc}
Audubon	436	3	19	0
Calhoun	562	3	22	0
Carroll	566	1	39	0
Crawford	709	6	50	0
Greene	566	4	23	0
Guthrie	586	2	34	0
Ida	432	2	20	0
Sac	573	4	44	1
Shelby	579	3	31	1
Total	5,009	28	282	2

TM based estimates of corn and soybean acreage were computed for all nine counties in the study area. Three counties (Calhoun, Crawford, and Ida) were not completely contained within the TM scene. Table 1 shows the total number of segments in the population (N_{hc}) and the number of sample segments (n_{hc}) in each county, by stratum. Synthetic estimation was used within stratum A for the parts of counties outside the scene, and in stratum B for all nine counties. Table 2 gives the computed county estimates by stratum and estimation method.

Table 2: Iowa 1988 County Estimates of Crop Acreage by Stratum and Estimation Method

County	Stratum A Battese-Fuller	Stratum A Synthetic	Stratum B Synthetic	Total	C.V.
Corn	acres (000)	acres (000)	acres (000)	acres (000)	percent
Audubon	91.9	-	.3	92.2	3.5
Calhoun	130.3	2.6	.4	133.2	2.9
Carroll	140.7	-	.7	141.4	3.2
Crawford	128.4	23.4	.9	152.7	3.1
Greene	129.6	-	.4	130.0	3.0
Guthrie	105.7	-	.6	106.3	4.9
Ida	43.4	63.2	.4	107.0	3.7
Sac	137.5	-	.8	138.3	2.9
Shelby	140.2	-	.5	140.7	2.9
Total	1047.7	89.2	5.0	1141.8	
Soybeans					
Audubon	69.8	-	.1	69.9	6.6
Calhoun	143.2	1.7	.1	145.0	4.0
Carroll	106.6	-	.1	106.7	9.0
Crawford	91.3	15.5	.2	106.9	5.4
Greene	117.4	-	.1	117.5	4.6
Guthrie	64.3	-	.1	64.4	10.9
Ida	34.6	41.7	.1	76.4	6.9
Sac	112.8	-	.1	112.9	4.9
Shelby	80.9	-	.1	81.0	7.4
Total	820.9	58.8	1.0	880.7	

Table 3 contains the official county estimates issued by the Iowa Agricultural Statistics Service. These published estimates are based on additional survey and administrative data (see Chapter 7), and are considered as the standard for evaluating the Battese-Fuller model values. The tables show that the computed county estimates for corn were more efficient overall than those for soybeans. For eight of the nine counties, the C.V. for corn was less than 4 percent. No county had a C.V. of less than 4 percent for soybeans. The percent difference ranged from 0.2 to 9.2 for corn, and from 0.8 to 17.8 for soybeans.

Table 3: County Estimates for 1988 Iowa Study

County	Corn			Soybeans		
	Official	Computed	% Diff.[a]	Official	Computed	% Diff.[a]
	acres (000)	acres (000)		acres (000)	acres (000)	
Audubon	100	92.2	7.8	70.7	69.9	1.1
Calhoun	133	133.2	0.2	150.0	145.0	3.3
Carroll	141	141.4	0.3	117.0	106.7	8.8
Crawford	147	152.7	3.9	106.0	106.9	0.8
Greene	125	130.0	4.0	143.0	117.5	17.8
Guthrie	98	106.3	8.5	77.5	64.4	16.9
Ida	112	107.0	4.5	75.2	76.4	1.6
Sac	136	138.3	1.7	124.0	112.9	9.0
Shelby	155	140.7	9.2	94.9	81.0	14.6
Total	1,147	1,141.8		958.3	880.7	

[a] $\%Diff. = \dfrac{|\,Official - Computed\,|}{Official} * 100$

6.5 Evaluation Practices

NASS first began to address the problem of applying satellite data to small area estimation in the mid 1970's. In 1976, Huddleston and Ray (1976) proposed that within each stratum, the mean pixels per segment calculated by classifying all segments within an entire analysis district be replaced by the mean pixels per segment computed by classifying all segments within a given county. This county pixel mean is substituted into the corresponding stratum regression equation for the crop of interest. Amis, Martin, McGuire, and Shen (1982) describe the Huddleston-Ray estimator as an analysis district regression estimator applied to a subarea of the analysis district. The regression coefficients are estimated from sampled segments located throughout the analysis district, while the mean being estimated is from a subpopulation of the analysis district. The Huddleston-Ray estimator is simple and intuitively appealing, but Walker and Sigman (1982) point out two major drawbacks. First, it is unclear how to accurately compute the variance of the estimator. Second, the estimator lumps together a term attributable to sampling error within a given county and another term that measures the inherent distinction between a county and the analysis district. Amis et al. (1982) empirically demonstrate that the Huddleston-Ray method can generate biased estimates and that the variance estimation formula can overestimate the variability for a given county. The mathematical formulas for the Huddleston-Ray estimator and its variance estimator are provided in Appendix B.

The problems with the Huddleston-Ray estimator documented by Walker and Sigman (1982) and by Amis et.al (1982) were recognized soon after its development and prompted Cardenas, Blanchard, and Craig (1978) to devise a different type of estimator. The Cardenas family of estimators has three forms, each of which uses auxiliary Landsat data through a regression type estimator. However, the versions use different methods of estimating the slope term. The three forms are the ratio estimator, the separate regression estimator, and the combined regression estimator. (Appendix C gives the mathematical formulation for the Cardenas family of estimators.) As with the Huddleston-Ray method, within each stratum the Cardenas method compares the analysis district level mean pixels per segment classified to a crop to the corresponding county level mean for that crop. However, the Cardenas methods uses all segments in the analysis district to calculate the analysis district mean, where the Huddleston-Ray approach only uses sample segments. The estimate of average crop area per segment is adjusted by an amount proportional to this difference between the county and analysis district means. Amis et al. (1982) examined the ratio and separate regression Cardenas estimators, and compared them with the Huddleston-Ray estimator. Cardenas et al. (1978) stated that none of the estimators they presented were shown to be "best" in any sense, nor did they demonstrate any optimum properties. They did show that each of these estimators, when summed over counties, provides an unbiased stratum level estimate for the State. Also, assuming that the within county variance is the same for all counties, the method enables unbiased estimation of the State-wide variance. Amis et al. (1982) emphasized that an unbiased estimate of the county mean crop area per segment may not be possible when there are few sample segments in a county. Whenever there are significant differences in county variances, the Cardenas estimators appear to have higher variances than the Huddleston-Ray estimator. Amis et al. (1982) concluded that there appears to be no difference between the Cardenas ratio estimator and the separate regression estimator, and that the Cardenas estimators do not perform better than the Huddleston-Ray estimator. Both Cardenas estimators studied appeared to be biased, with larger variances than the Huddleston-Ray estimator.

The Cardenas method was never used in an operational remote sensing program since it did not provide sufficient improvement over the Huddleston-Ray estimator. The Huddleston-Ray estimator was used to generate county estimates for use by the NASS State Statistical Offices (SSO's) until 1982. At that time, Walker and Sigman (1982) advised that calculation of county estimates using the Huddleston-Ray method be discontinued, and that the Battese-Fuller method be used instead.

Walker and Sigman (1982) studied the Battese-Fuller model using Landsat MSS data over a six county region in eastern South Dakota. They found a modest lack of fit of the model, with larger model departure corresponding to low correlation between classified pixel counts and ground survey observations. A key feature of the Battese-Fuller model is the county effect parameter and this effect was found to be highly significant for corn, the most prevalent of the four crops considered in the study. Furthermore, this effect manifested itself within several strata but was negligible across strata. The study nonetheless indicated robustness of the Battese-Fuller estimators against departure from certain model assumptions. Two members of the Battese-Fuller family satisfied the criterion for small relative root mean square error; i.e. less than 20 percent of the estimate was attributable to root mean square error. These members were the estimators that minimized mean square error and bias, respectively, under the model assumptions. However, the Battese-Fuller estimate closest to the Huddleston-Ray estimate was far less satisfactory, failing to meet the desired upper limits for mean square error and bias. This study provided the justification for replacing the Huddleston-Ray estimator with the Battese-Fuller family.

The reliability of the county estimates based on the Battese-Fuller model has been closely watched since its implementation in 1982. As mentioned previously, these estimates are only one of possibly four or more indications that are composited to provide the final published crop acreage values. The reliability of the Battese-Fuller estimates can vary between years, between crops, between counties and between States depending on the stage of the crop at the time of the Landsat imagery, the amount of crop acreage within the county, the number of segments within the county, and cloud cover. The results presented in Tables 2 and 3 for corn are relatively good with all CVs less than 5 percent and over half of the percentage differences from the published value less than 4 percent. The soybean results are slightly poorer, with CVs ranging from 4 percent to 11 percent and percentage differences ranging up to 18 percent. Table 4 presents more recent results for a set of counties covered by Landsat in Mississippi for 1991. A review of the CVs and percentage differences indicate that the Battese-Fuller estimates can have relatively large CVs and percentage differences when the county crop acreage is less than 30,000 acres. Some summary statistics of the differences for the four crop examples discussed are presented in Table 5. The mean average difference is typically less than 10,000 acres, but for small county acreages such as rice in Mississippi, large percentage differences may still occur. Consequently, NASS SSO's still use additional survey and administrative data to help set the published values.

Table 4: County Estimates for Mississippi 1991

County	Official	Computed	% Diff[a]	CV
Cotton	acres (000)	acres (000)	percent	percent
Bolivar	65.5	61.6	6.0	9.9
Coahoma	105.7	88.3	16.5	4.8
Humphreys	61.6	57.3	7.0	5.9
Issaquena	38.0	34.6	9.0	11.3
Leflore	79.2	87.8	10.9	4.0
Quitman	31.0	46.4	49.7	8.6
Sharkey	47.0	48.6	3.4	7.0
Sunflower	100.0	79.3	20.7	6.9
Tallahatchie	64.2	67.9	5.8	7.2
Tunica	45.6	38.0	16.7	6.6
Washington	95.7	102.4	7.0	3.9
Yazoo	94.5	93.9	.6	8.0
Total	828.0	806.1		
Rice				
Bolivar	74.0	66.2	10.5	5.4
Coahoma	15.8	10.4	34.2	24.0
Humphreys	3.6	7.1	97.2	32.4
Leflore	16.6	19.4	16.9	18.6
Sharkey	5.0	7.8	56.0	21.8
Sunflower	36.0	37.8	5.0	9.3
Tallahatchie	9.6	8.5	11.5	35.3
Tunica	17.5	9.9	43.4	26.3
Washington	30.5	22.6	25.9	15.5
Total	208.6	189.7		

[a] $\%Diff. = \dfrac{|\, Official - Computed \,|}{Official} * 100$

Table 5: Summary Statistics on Accuracy of Battese-Fuller Estimates
(1000 acres)

Crop/State/Year	MD*	RMSD*	MAD*	LAD*
Corn Iowa 1988	-0.6	6.8	5.4	14.3
Soybeans Iowa 1988	-8.6	11.9	9.1	25.5
Cotton Mississippi 1991	-1.8	10.0	7.8	20.7
Rice Mississippi 1991	-2.1	5.2	4.5	7.9

*   MD   =   mean difference between Battese-Fuller and published value
    RMSD =   root mean squared deviation
    MAD   =   mean absolute difference
    LAD   =   largest absolute difference

6.6 Current Problems and Activities

As technology improves, new sensors produce satellite data that can be more accurately classified to a given crop than ever before. Consequently, the overall count of pixels classified to a given crop within a county can possibly be used directly to estimate crop acreage. The overall pixel count represents a census of pixels covering the county and therefore is not subject to sampling error. However, a nonsampling error is introduced due to inaccuracies in the classification. A general expression for such an estimator is:

$$\hat{T}_c = \eta X_c$$

where:

X_c =   number of pixels classified to crop of interest in county c

η =   adjustment term

Two adjustment terms that are being investigated are:

1) a conversion factor representing the area on the ground corresponding to one pixel for the specific sensor used, and

2) a combined ratio of the form:

$$\frac{\sum\limits_{h} N_h \bar{y}_{h..}}{\sum\limits_{h} N_h \bar{x}_{h..}}$$

where:

N_h $\quad = \quad$ number of segments in stratum h

$\bar{y}_{h..}$ $\quad = \quad$ mean reported crop area per segment in stratum h

$\bar{x}_{h..}$ $\quad = \quad$ mean number pixels per segment classified to the crop in stratum h.

Both adjustment terms are conceptually simple. The combined ratio uses stratum level survey information to compute the adjustment term that may provide a more accurate conversion of pixel counts to crop area than the set conversion factor. Also, the ratio has a readily available formula for estimating the variance.

Research continues to focus on identifying new geographic areas and crops where this estimator would be applicable. Also, possible benefits of remotely sensed data from alternative sources, such as radar satellites, will be investigated as the newer sources are available. In recent years TM sensor data have been used to produce county estimates in the Delta region. County estimates of rice, cotton, and soybeans were produced for Arkansas and Mississippi in 1991, with Louisiana added in 1992. In 1993 satellite data were only used in Arkansas due to budgetary constraints. To date, the satellite based estimates have only been produced on a limited scale. The NASS SSO's continue to rely on other data series for helping set the published county estimates of crop acreages. They conduct a large non-probability county estimates survey (see Chapter 7) that serves a dual purpose of also providing updated control data for the list sampling frame. This is an integral part of the NASS survey program and so will continue in some form for the foreseeable future. Fairly reliable administrative data sources are also available. NASS is continuing to investigate the benefits of satellite based county estimates in relation to these other available data sources. One by-product of the satellite data process that is attractive to the State offices is color coded land use maps at the county level. These maps provide a pictorial view of the distribution of the crops within each county. Identifying alternative uses of satellite data such as this is an important research objective of NASS.

REFERENCES

Allen, J.D. (1990), "A Look at the Remote Sensing Applications Program of the National Agricultural Statistics Service," Journal of Official Statistics, 6, pp. 393-409.

Amis, M.L., Martin, M.V., McGuire, W.G., and Shen, S.S. (1982), "Evaluation of Small Area Crop Estimation Techniques Using Landsat and Ground-Derived Data," LEMSCO-17597, Houston, TX: Lockhead Engineering and Management Services Company, Inc.

Battese, G.E., Harter, R.M., and Fuller, W.A. (1988), "An Error-Components Model for Prediction of County Crop Areas Using Survey and Satellite Data," Journal of the American Statistical Association, 83, pp. 28-36.

Bellow, M.E. (1991), "Comparison of Sensors for Corn and Soybean Planted Area Estimation," NASS Staff Report No. SRB-91-02, U.S. Department of Agriculture.

Bellow, M.E. and Graham, M.L. (1992), "Improved Crop Area Estimation in the Mississippi Delta Region using Landsat TM Data," Proceedings of the ASPRS/ACSM Convention, Washington, D.C., pp. 423-432.

Cardenas, M., Blanchard, M.M., and Craig, M.E. (1978), "On The Development of Small Area Estimators Using LANDSAT Data as Auxiliary Information," Economic, Statistics, and Cooperatives Service, U.S. Department of Agriculture.

Cochran, W.G. (1977), "Sampling Techniques," New York, N.Y.: John Wiley & Sons.

Fuller, W.A. and Battese, G.E. (1973), "Transformations for Estimation of Linear Models with Nested-Error Structure," Journal of the American Statistical Association, 68, pp. 626-632.

Huddleston, H.F. and Ray, R. (1976), "A New Approach to Small Area Crop Acreage Estimation," Proceedings of the Annual Meeting of the American Agricultural Economics Association, State College, PA.

Ozga, M., Mason, W., and Craig, M. (1992), "PEDITOR - Current Status and Improvements," Proceedings of the ASPRS/ASCM Convention, Washington, D.C., pp. 175-183.

Prasad, N.G.N. and Rao, J.N.K. (1990), "The Estimation of the Mean Squared Error of Small-Area Estimators," Journal of the American Statistical Association, 85, pp. 163-171.

Walker, G. and Sigman, R. (1982), "The Use of LANDSAT for County Estimates of Crop Areas - Evaluation of the Huddleston-Ray and Battese-Fuller Estimators," SRS Staff Report No. AGES 820909, U.S. Department of Agriculture.

Appendix A: Estimators of Battese-Fuller Variance Components

The estimators of the Battese-Fuller variance components, σ^2_{vh} and σ^2_{eh}, at the analysis district level are expressed as:

$$\hat{\sigma}^2_{eh} = [1/(n_h-C-1)] \sum_{c=1}^{C} \sum_{i=1}^{n_{hc}} [y_{hci}-\bar{y}_{hc.}-\hat{\alpha}_h(x_{hci}-\bar{x}_{hc.})]^2$$

$$\hat{\sigma}^2_{vh} = \max([s^2_{uh}-(n_h-2)\hat{\sigma}^2_{eh}]/(n_h-T_h),0)$$

where:

$$\hat{\alpha}_h = \frac{\displaystyle\sum_{c=1}^{C} \sum_{i=1}^{n_{hc}} (x_{hci}-\bar{x}_{hc.})(y_{hci}-\bar{y}_{hc.})}{\displaystyle\sum_{c=1}^{C} \sum_{i=1}^{n_{hc}} (x_{hci}-\bar{x}_{hc.})^2}$$

$$s^2_{uh} = \sum_{c=1}^{C} \sum_{i=1}^{n_{hc}} (y_{hci}-\hat{\beta}_{0h}-\hat{\beta}_{1h}x_{hci})^2$$

$$T_h = \frac{n_h\displaystyle\sum_{c=1}^{C} n^2_{hc}\bar{x}^2_{hc.} + (\displaystyle\sum_{c=1}^{C} n^2_{hc})(\displaystyle\sum_{c=1}^{C} \sum_{i=1}^{n_{hc}} x^2_{hci}) - 2n_h\bar{x}_{h..}\displaystyle\sum_{c=1}^{C} n^2_{hc}\bar{x}_{hc.}}{(n_h\displaystyle\sum_{c=1}^{C} \sum_{i=1}^{n_{hc}} x^2_{hci}) - n^2_h\bar{x}^2_{h..}}$$

n_h	=	number of sample segments in stratum h
n_{hc}	=	number of sample segments in stratum h, county c
C	=	number of counties in analysis district
y_{hci}	=	reported acres of crop of interest in stratum h, county c, segment i

$\bar{y}_{hc}$ =   mean reported acres of crop of interest per sample segment in stratum h, county c

x_{hci} =   number of pixels classified to crop of interest in stratum h, county c, segment i

$\bar{x}_{hc.}$ =   mean number of pixels classified to crop of interest per sample segment in stratum h, county c

$\bar{x}_{h..}$ =   mean number of pixels classified to crop of interest per sample segment in stratum h

$\hat{\beta}_{oh}, \hat{\beta}_{1h}$ =   least squares regression parameters in stratum h for regression of y_{hci} on x_{hci} .

Appendix B: Huddleston-Ray Estimator

The Huddleston-Ray estimator replaces the classified pixel average for the analysis district with the classified pixel average for a county when estimating the county mean crop area per frame unit. Within the analysis district, the overall mean crop area in regression stratum h is estimated by:

$$\bar{y}_{(REG),h..} = \bar{y}_{h..} + \hat{\beta}_h (\bar{X}_{h..} - \bar{x}_{h..})$$

and the stratum level mean crop area for county c is estimated by:

$$\bar{y}_{(HR),hc.} = \bar{y}_{h..} + \hat{\beta}_h (\bar{X}_{hc.} - \bar{x}_{h..})$$

where:

$\bar{y}_{h..}$ =   mean reported area per sample segment for crop of interest in stratum h

$\bar{x}_{h..}$ =   mean number of pixels per sample segment classified to crop of interest in stratum h

$\bar{X}_{h..}$ =   mean number of pixels per segment classified to crop of interest in stratum h

$\bar{X}_{hc.}$ =   mean number of pixels per segment classified to crop of interest in stratum h, county c

β_h =   least squares regression slope parameter estimate in stratum h for

regression of y_{hci} on x_{hci} .

The Huddleston-Ray estimator of total crop area in the regression strata of county c
is then:

$$\hat{T}_{(HR),c} = \sum_{h=1}^{H_r} N_{hc}[\bar{y}_{h..} + \hat{\beta}_h(\bar{X}_{hc.} - \bar{x}_{h..})]$$

where:

N_{hc} =   number of segments in stratum h, county c

H_r =   number of regression strata.

Walker and Sigman (1982) pointed out two problems with the Huddleston-Ray
estimator. They stated that the variance calculation was unclear, and use of the
difference:

$$\bar{X}_{hc.} - \bar{x}_{h..} = (\bar{X}_{hc.} - \bar{x}_{hc.}) + (\bar{x}_{hc.} - \bar{x}_{h..})$$

combined within-county sampling error with the county analysis district difference.
However, Amis et al. (1982) gave the following estimator for the variance:

$$V(\hat{T}_{(HR),c}) = \sum_{h=1}^{H_r} N^2_{hc} \frac{N_h - n_h}{n_h} s_h^2 \frac{n_h-1}{n_h-2}(1-r_h^2)T_{hc}$$

where:

N_h =   number of segments in stratum h

n_h =   number of sample segments in stratum h

x_{hci} =   number of pixels classified to crop of interest in stratum h, county
c, segment i

s_h^2 =   sample variance for the reported area in stratum h

= $\sum_{c=1}^{C} \sum_{i=1}^{n_h} \frac{(y_{hci} - \bar{y}_{h..})^2}{n_h - 1}$

r_h^2 = sample coefficient of determination for stratum h

C = number of counties in analysis district

$$T_{hc} = \frac{1}{n_h} + \frac{(\bar{X}_{hc.}-\bar{x}_{h..})^2}{\displaystyle\sum_{c=1}^{C}\sum_{i=1}^{n_h}(x_{hci}-\bar{x}_{h..})^2} \quad .$$

Appendix C: Cardenas Family of Estimators

The Cardenas family of estimators uses the stratum level differences between mean number of pixels classified to the crop of interest in the county and the analysis district, respectively, to adjust the mean reported crop area per sample segment. Within a regression stratum h, the estimate of mean crop area per segment for a county c is:

$$\bar{y}_{(CAR),hc.} = \bar{y}_{h..} + \hat{\beta}_h(\bar{X}_{hc.}-\bar{X}_{h..})$$

where:

$\bar{X}_{hc.}$ = mean number of pixels per segment classified to crop of interest in stratum h, county c

$\bar{X}_{h..}$ = mean number of pixels per segment classified to crop of interest in stratum h.

(See Appendix B for definitions of all terms not defined here). The estimate of total crop area in the regression strata of county c is:

$$\hat{T}_{(CAR),c} = \sum_{h=1}^{H_r} N_{hc}[\bar{y}_{h..} + \hat{\beta}_h(\bar{X}_{hc.}-\bar{X}_{h..})] \quad .$$

The parameter $\hat{\beta}_h$ relates classified pixel counts to reported crop area. Cardenas

et al. (1978) proposed three alternative estimators for $\hat{\beta}_h$:

(1)   Ratio estimator

$$\hat{\beta}_h = \bar{y}_{h..}/\bar{X}_{h..}$$

(2)   Separate regression estimator

$$\hat{\beta}_{lh} = \frac{N_h \sum_{c=1}^{C} n_{hc}(\bar{X}_{hc.} - \bar{X}_{h..})\bar{y}_{hc.}}{n_h \sum_{c=1}^{C} N_{hc}(\bar{X}_{hc.} - \bar{X}_{h..})^2}$$

(3)   Combined regression estimator

$$\hat{\beta}_h = \frac{\sum_{h=1}^{H_r} (N^2_h/n_h) \sum_{c=1}^{C} n_{hc}(\bar{X}_{hc.} - \bar{X}_{h..})\bar{y}_{hc.}}{\sum_{h=1}^{H_r} N_h \sum_{c=1}^{C} N_{hc}(\bar{X}_{hc.} - \bar{X}_{h..})^2}.$$

The combined regression estimator is applicable only when the $\hat{\beta}_h$'s are assumed to be constant over strata.   Cardenas et al. (1978) also provided formulas for estimating the variance of their county estimators.

CHAPTER 7

The National Agricultural Statistics Service County Estimates Program

William C. Iwig
National Agricultural Statistics Service

7.1 Introduction and Program History

The National Agricultural Statistics Service (NASS) of the U.S. Department of Agriculture (USDA) publishes over 300 reports annually regarding the Nation's crop acreage, crop production, livestock inventory, commodity prices, and farm expenses. The primary source of this information is surveys of U.S. farmers, ranchers, and agribusinesses who voluntarily provide information on a confidential basis. These surveys are normally designed to provide State and U.S. level indications of agricultural commodities. There is also a need for county level estimates to assist farmers, ranchers, agribusinesses, and government agencies in local agricultural decision making.

NASS has published annual county estimates for over 70 years through funding provided by cooperative agreements with State departments of agriculture and agricultural universities, and directly from other USDA agencies. The earliest known record of published county estimates is by the Wisconsin State Board of Agriculture, which issued county estimates on acreage and production of crops for 1911 and 1912 along with the number and value of livestock for 1912. Not until 1917, following the signing of the first Federal-State cooperative agreement, did the USDA assist in the preparation and publication of the Wisconsin county estimates. The cooperative agreement helped eliminate duplication of efforts between Federal and State statisticians, making possible more service for less cost. The cooperative work grew rapidly after 1917 as other State departments of agriculture and State agricultural universities established cooperative agreements with the USDA. State governments needed county level information and their funding made possible the publication of county level estimates by USDA.

The New Deal Farm Programs of President Franklin D. Roosevelt's Administration used county estimates of agricultural commodities extensively and refocused USDA's attention to these estimates. In May 1933, the Agricultural Adjustment Act was passed and the Agricultural Adjustment Administration (AAA) was soon in place. This agency had the task of reducing supply in order to improve prices of agricultural commodities. These programs greatly increased demands on NASS for county estimates of commodities used by the AAA to set county quotas and program pay-outs for surplus items.

In more recent years, the Federal Crop Insurance Corporation (FCIC) and the Agricultural Stabilization and Conservation Service (ASCS) of the USDA have used NASS county estimates to administer their programs and they provide funding to NASS for that purpose. Their programs involve payments to farmers if crop yields are below certain levels. Both agencies have chosen to use the NASS county estimates, when available, as the basis for determining these payments.

The estimation approach has remained relatively unchanged over the years. The basic process for estimating totals such as crop acreage and livestock inventory initially involves scaling various survey estimates and other available administrative data at the county level to be additive to the official USDA State level estimate. These scaled estimates are composited together, usually with the previous year estimate, to provide the actual county estimate for the current year. This scaling and compositing process tends to strengthen the final estimate over a direct design based expansion. These estimates are checked against any available administrative data that are reliable indicators of minimum levels and modifications are made if necessary. Program changes that have been made since 1917 involve data processing advances, allowing more data to be used, and larger sampling frames and more sophisticated sample selection techniques, providing better coverage of the farm population. Also, advances have been made to improve the quality of the State level estimates, which indirectly benefit the quality of the published county estimates through the scaling process. In the late 1950's, methodology was developed to conduct probability area frame surveys, where random segments of land would be selected for enumeration. In the 1960's these surveys became operational, which provided for the first time probability survey indications of crop acreage and livestock inventories on a State level basis. During this time frame, the State reporter lists were also increasing in size and improving in quality. With improved data processing capabilities in the 1970's, probability Multiple Frame (MF) Surveys were implemented at the U.S. and State levels, which combined the use of list and area sampling frames. Also, some States have conducted probability or quasi-probability MF County Estimates surveys (North Carolina Ag Statistics Service 1986).

States have traditionally shown a large degree of autonomy in designing and conducting their county estimates surveys. This has been due, in large part, to funding from the State cooperator, the quality of different data sources and different computing capabilities in each State. In the late 1980's, a NASS task force developed a County Estimates System for sample selection and summarization that provides a general framework, but still allows considerable flexibility to each State in their sample selection and summarization procedures (Bass et al. 1989). This system is now the standard being used by NASS State offices for their county estimates program.

7.2 Program Description, Policies, And Practices

The NASS County Estimates Program is really 45 different programs conducted separately by each NASS State Statistical Office (SSO). There is some general structure provided by the 1989 County Estimates Task Group, but still each State has considerable flexibility in the implementation of the procedures. The quality of the county estimates is to some degree related to the amount of financial support being provided by the State cooperator, which is usually the State Department of Agriculture.

The U.S. Census of Agriculture, conducted by the Bureau of the Census, has always served as a benchmark for the USDA crop and livestock estimates, and especially for county estimates. Since 1920 the Census has been conducted every five years (on a 4 year schedule from 1974 to 1982), providing county, district, State, and U.S. level estimates of most agricultural commodities. Since 1982 the Census has been conducted to coincide with the economic censuses (business, industry, etc.) in years ending in 2 and 7. Census county level estimates are closely watched since the USDA estimates are often based on very few survey returns. At the same time, the quality of the Census numbers are also closely evaluated. The completeness of the Census varies from State to State, county to county, and item to item. Consequently, the Census values are interpreted differently. After the Census values are published, NASS statisticians review their estimates and make revisions as necessary.

Another major component to the County Estimates Program has been the official USDA State level estimate. Preliminary survey estimates and administrative data are scaled to be additive to the official State total. State estimates are based on more data than each individual county estimate and, in recent years, have been based on probability survey indications. Consequently, the State estimates have always been considered more reliable than any individual county estimate. In addition to being more reliable, State level estimates are usually already published before county estimates are published. For these reasons, county level indications have always been scaled to the State level estimates rather than the State level estimate being the sum of independently derived county estimates.

Over the years, the county estimate surveys have developed into a major source of information for list frame maintenance and updating. Farm operations that had not been contacted within a prescribed time frame can be targeted for sampling for the annual county estimates survey. Currently, NASS has a stated policy that all control data on the list sampling frame (LSF) should be less than five years old (USDA 1991, Policy and Standards Memorandum 14-91). Control data refers to the historic survey data values or data values from external sources that are stored on the LSF and used for stratification and sample design purposes.

Another policy that is followed in all States is the suppression of any county estimate that would disclose the data of any individual operation, as specified in Policy and Standards Memorandum 12-89 (USDA 1989). This policy preserves the confidentiality of all reports, which is a foundation of voluntary reporting to NASS. Estimates cannot be published if either: (1) the estimate is based on information from fewer than three respondents, or (2) the data for one respondent represents more than 60 percent of the estimate. Exceptions to this rule are only granted when written and signed permission is given by the respondent. Suppressed estimates may be combined with another county as long as the confidential data are not disclosed.

In most States, county estimates are made for all major crop and livestock categories. This may cover 50 to 100 separate commodity items. Estimates for crop items usually include planted acres, harvested acres, yield, production, and value of production for a particular crop year. Some States also publish separate estimates for different cropping practices, such as irrigated and non-irrigated acreages. Livestock estimates include inventory numbers on a particular date, possibly marketings, and inventory value. Each SSO develops their own county estimate publication because they are State funded. These estimates have associated sampling and non-sampling errors. No variances or error information are published for the final county estimates. Mean squared error information is only published for major agricultural items at the U.S. level.

7.3 Estimator Documentation

The NASS County Estimates System uses a combination of scaling and compositing techniques to provide a county level total estimate for any particular agricultural item. Separate estimates that may be composited together include the previous year official estimate, current year direct expansion and ratio estimates, and other available indications. In recent years, remotely sensed data from satellites have been used to generate county level estimates of crop acreages for selected crops where this technology has been applied (see Chapter 6). County estimates of a ratio such as crop yield, which is the ratio of total crop production to total harvested acres, are dependent on the final estimates of the two items involved. Current year data are collected using primarily a mail survey in the fall of the year with some selected telephone follow-up. State sample sizes can range up to 40,000 with usable record counts around 200 for major items in major counties. However, county estimates for many commodities are based on fewer than 20 sample records.

A key feature of the system is the sample design which involves selecting sampling units from multiple overlapping stratified designs. A separate design is developed for each commodity of interest. The system combines data collected from sampled operations from these different designs such that the selection probabilities are not used in calculating the survey estimates. Another key feature of the system is the coordination of survey contacts from the different designs to control respondent burden. A third feature is a synthetic scaling of the county estimates in order that they sum to the official U.S. Department of Agriculture State level estimates. A fourth feature is the compositing of the different estimates to provide final county level estimates. Further details on each of these features follow.

7.3.1 Commodity Specific Stratified Designs

The NASS County Estimates Program depends primarily on a large mail survey in the fall of the year with State level sample sizes ranging up to 40,000. Some States conduct two surveys, with an early fall survey covering acreage and production of small grains which are usually harvested by September. Then the late fall survey covers the fall harvested crops and livestock. The sample units are farm operations selected from the NASS list sampling frame in each State.

One of the major goals of the County Estimates System is to provide a framework that will ensure adequate representation for each agricultural item of interest. In order to provide adequate county level estimates, major farm operations for each item of interest must be represented appropriately in the sample. This is relatively easy for the major crops in a State since a sample design representing all known operations with cropland would represent any major crop adequately. However, in order to provide adequate representation for rare crop and livestock items, separate stratified sample designs are developed for each agricultural commodity as needed. The sample design strata for each commodity are based on the positive control data for that particular item. Control data are the historic data values stored on the list sampling frame. Strata boundaries typically coincide with the categories used in the Census of Agriculture publications. Table 1 illustrates the stratified design that might be developed for barley in a particular State, covering all known operations that have positive control data for barley.

The major function of the stratified design is to provide a framework to group similar size operations for summarization (see 7.3.3). Initial sampling may occur at the State level within each stratum. Or, different sampling rates may be used at the county level in order to assure an adequate sample within each county. Different sampling rates by county would typically occur when the commodity frame contains only a few records in a particular county. It may be necessary to sample all records with "probability one" in that county, where a smaller sampling fraction is sufficient in other counties. This most frequently occurs with rare commodities. Another sampling option keys on whether the sampling unit reported in the previous year. If the current to previous year ratio is a primary indication for a State, units that reported in the previous year may be sampled heavily, and other records sampled at a lighter rate.

Table 1. Example Stratified Design for Barley

Stratum	Population Count	Boundary (acres)
10	2,500	1 - 49
20	1,000	50 - 99
30	400	100 - 299
40	100	300+
Total	4,000	

7.3.2 Coordination of Multiple Samples

The samples selected from the different commodity designs contain many overlapping records. A farming operation could easily be selected from multiple commodity designs. In addition, many of the selected operations may have already provided all or some of the requested information on another current year survey. These other survey data files are used as input to the County Estimate System. The system is designed to identify which records already have provided the requested information and questionnaires are not sent to these operations. Even if an operation has only provided some of the needed data on previous crop specific or livestock specific surveys, it will typically not be recontacted to help control respondent burden. Data items not included on the previous surveys are treated as "missing" in the county estimates expansions. The system also identifies which records are duplicated in multiple designs and in multiple samples. Only one questionnaire is sent to each sampled unit. The same questionnaire, containing all items of interest, is used regardless of the commodity design (barley, corn, hogs, etc.) from which the record was selected. There is usually some telephone follow-up to non-respondents as resources allow. Telephoning may be targeted to provide sufficient data for each commodity. Since a secondary objective of the county estimate survey is to update control data on the list sampling frame, some telephoning may be targeted at operators with missing control data or control data that are more than five years old.

7.3.3 Creation of Survey Indications

The County Estimates System is designed to provide direct expansion and ratio estimates based on sample data collected from the county estimates survey and from sample data collected from other current year surveys. As mentioned previously, the same questionnaire is used for all farm operations selected specifically for the county estimates survey, regardless of the originating commodity design. Consequently, a farm operation selected from the barley design will also be asked to provide data on all other crop and livestock items. All reported data from the county estimates survey and from other surveys are used in providing the survey indications. For each operation, the system identifies the assigned strata from all of the commodity designs. All records will not be included in each commodity design since all records do not have positive control data for all commodities. Records that do not have an original design stratum for a commodity are assigned to "pseudo stratum 99" for summary. Then corn data are summarized in the corresponding stratum from the corn design for each operation and hog data are summarized in the corresponding hog stratum. Since data are used for a particular item from records that were <u>not</u> selected in the original sample design, the direct expansion and ratio estimates are not based on the selection probabilities. However, this approach probably doubles the number of positive data records available for most survey items compared to just using data records from the original commodity designs. The use of this additional data is a stabilizing factor in providing reliable county level estimates.

Survey estimates from the County Estimates System are provided at State, district, and county levels for each item. Districts are groups of geographically contiguous counties with relatively homogeneous agricultural practices and climate within each district. There are usually four to nine districts per State. The State and district estimates are used primarily in the scaling process described later. The county level survey estimates are the basis for the final published estimates, but they also go through a scaling and compositing process. Population counts and useable record counts are generated by the system at each level. The direct expansion estimate for a particular commodity at any level is represented as follows:

$$\hat{T}_{(E),d} = \sum_{h=1}^{H} \frac{N_{dh}}{n_{dh}} \sum_{i=1}^{n_{dh}} y_{dhi}$$

where:

d   =   domain indicator (State, district, or county)

$\hat{T}_{(E),d}$ =   direct expansion estimate for domain d

N_{dh} =   population count for stratum h, domain d

n_{dh} =   number of usable records for stratum h, domain d

y_{dhi} =   reported value for i^{th} record in stratum h, domain d.

Expansion factors (N_{dh}/n_{dh}) are generated for each stratum within each design at the county and district levels as if the sampling occurred at those levels. The n_{dh} refers to the number of usable records in stratum h, domain d, which includes records from multiple sampling designs. Consequently, the quantity N_{dh}/n_{dh} does not represent the actual sampling weight of any survey record. Under this approach, county level estimates are not necessarily additive to the district and district level estimates are not necessarily additive to the State. Table 2 provides an illustration for an example stratum in a State with four counties and two districts (USDA 1992). The county expansions do not add to the district nor do the district expansions add to the State at this stage. In addition, the county, district, and State estimates will typically all be

Table 2: Examples of Direct Expansion County, District, and State Estimates, for Corn Planted Acres

Stratum	District	County	N_{dh}	n_{dh}	N_{dh}/n_{dh}	$\sum_{i=1}^{n_{dh}} y_{dhi}$	$\hat{T}_{(E),d}$
01	10	003	25	10	2.50	400	1,000
01	10	005	300	30	10.00	700	7,000
01	10	999[a]	325	40	8.13	1,100	8,943
01	20	001	200	20	10.00	200	2,000
01	20	007	75	10	7.50	300	2,250
01	20	999[b]	275	30	9.17	500	4,585
01	99	999[c]	600	70	8.57	1,600	13,712

[a]District 10 values
[b]District 20 values
[c]State level values

biased downward due to the incompleteness of the list. For most major items, the separate commodity designs will only provide about 80% coverage. So scaling to the official USDA State estimate is absolutely necessary.

In addition to direct expansion estimates, ratio estimates of totals and ratio estimates of ratios are also created. For crop acreage items, possible ratio estimates are based on ratios of current year planted acres to previous year planted acres, harvested to planted acres, planted acres to total cropland acres, and irrigated acres to planted acres. The ratio estimates are generated from usable reports for both the numerator and denominator and are expressed as:

$$\hat{T}_{(R),d} = \left(\frac{\displaystyle\sum_{h=1}^{H} \frac{N_{dh}}{r_{dh}} \sum_{i=1}^{r_{dh}} y_{dhi}}{\displaystyle\sum_{h=1}^{H} \frac{N_{dh}}{r_{dh}} \sum_{i=1}^{r_{dh}} x_{dhi}} \right) X_d$$

$$= \hat{R}_{(R),d} X_d$$

where:

$\hat{T}_{(R),d}$	=	ratio estimate for domain d (State, district, or county)
N_{dh}	=	population count for stratum h, domain d
r_{dh}	=	number of usable reports for ratio in stratum h, domain d
y_{dhi}	=	reported value for i^{th} record in stratum h, domain d
x_{dhi}	=	reported value of auxiliary variable for i^{th} record in stratum h, domain d
X_d	=	value of auxiliary variable for domain d.

Unweighted ratios, calculated without the N_{dh}/r_{dh} stratum weight, may also be used. The actual ratio estimates generated are at the option of the State, dependent on their historic data series and what estimates the State has found to be reliable. The domain value of the auxiliary variable (X_d) is usually not a survey estimate, but a composite estimate based on multiple surveys and other indications (see 7.3.5).

The ratio estimate of the ratio, $\hat{R}_{(R),d}$, provides the initial estimate of such items as crop yield at the county level. If the number of reports is minimal and the estimate is not reasonable, the State office may adjust the estimate based on surrounding counties. For crop yield, these initial estimates are then typically multiplied by the final harvested acres estimates to provide estimates of total crop production. The production estimates are scaled to the official USDA State level estimate. This may necessitate further adjustments to the yield estimates. Consequently, the final yield estimates are actually derived from the county level harvested acres estimates and the State level production estimate. The scaling and compositing processes described in the next two sections only apply to yield estimates through their application to the production and harvested acres items.

7.3.4 Scaling of Indications

The first step in the process is to scale the individual county and district "indications" to the official published USDA State level estimate. Typically, "indications" that are scaled include:

    1)   survey direct expansion estimate
    2)   survey ratio estimates
    3)   previous year estimate
    4)   other indications (remotely sensed acreage estimates, Census of Agriculture, other Administrative data).

Initially, each district indication (direct expansion, ratio, administrative data) is scaled. Suppose there are "M" different indications. The scaling at the district level occurs as follows:

$$\hat{T}_{(SC_m),e} = \frac{\hat{T}_{(m),e}}{\displaystyle\sum_{e=1}^{E} \hat{T}_{(m),e}}\ \hat{T}_{(O),s}$$

where:

e	=	district index (e=1,..., E)
m	=	indication index (m=1,...,M)
$\hat{T}_{(m),e}$	=	value of mth indication, district e
$\hat{T}_{(O),s}$	=	official USDA level estimate, State s
$\hat{T}_{(SC_m),e}$	=	scaled estimate for m^{th} indication, district e.

The resulting district level estimates for each of the "M" indications, $\hat{T}_{(SC_m),e}$ will then sum to the official State estimate. Then, each of the "M" county level indications are scaled to the corresponding scaled district estimate as follows:

$$\hat{T}_{(SC_m),c} = \frac{\hat{T}_{(m),c}}{\sum_{c=1}^{C} \hat{T}_{(m),c}} \hat{T}_{(SC_m),e}$$

where:

c	=	county index within district (c=1,...,C)
$\hat{T}_{(m),c}$	=	value of mth indication, county c
$\hat{T}_{(SC_m),c}$	=	scaled estimate for m^{th} indication, county c.

The resulting county level estimates for each of the "M" indications (direct expansion, ratio, administrative data) then sum to the district estimate. This scaling process serves as a weighting adjustment to account for any incompleteness in the various indications. As mentioned previously, the NASS list sampling frame typically provides about 80% coverage for major commodities. Administrative data values also have varying degrees of completeness.

7.3.5 Compositing of Scaled Estimates

The next step in the process is to composite together the various scaled estimates to provide satisfactory county and district level estimates. The composite estimates generated for each county and district are represented as follows:

$$\hat{T}_{(COMP),d} = \sum_{m=1}^{M} W_{(m),d} \; \hat{T}_{(SC_m),d}$$

where:

d	=	domain indicator (district or county)
$W_{(m),d}$	=	composite weight for m^{th} indication, domain d
$\hat{T}_{(SC_m),d}$	=	m^{th} scaled estimate, domain d
$\hat{T}_{(COMP),d}$	=	composite estimate, domain d.

Currently, the composite weights are subjectively set by the statisticians in the State office to provide satisfactory and reliable estimates. They are subject to the conditions that they are non-negative and $\sum_{m=1}^{M} W_{(m),d} = 1$ for each domain. These weights are generally the same for a particular item in all counties and districts, but can be different when unusual survey data (outliers) cause a certain estimate to be unreliable. Typically, the previous year estimate is given some weight, which helps stabilize the composited value. This compositing is a form of an indirect estimator across time. The composite estimate borrows strength from previous estimates for the same small area domain.

Rounding rules are incorporated into this process so that the final estimates are the published values. These estimates are reviewed by statisticians in the State office for reasonableness based on their knowledge of the location and general size of the largest operations in the State for each commodity. The estimates must exceed minimum levels and not exceed maximum levels provided by reliable administrative data sources. For example, a State may check that the sum of major crop acreages does not exceed the Census of Agriculture total cropland acres for each county. If estimates are not reasonable, the data will be more closely examined for outliers and insufficient sample sizes. Different weights for the compositing process or adjustments to the outlier indications may be needed to provide the final published county level estimates.

7.4 Evaluation Practices

Each NASS State Statistical Office has taken a major responsibility in developing and evaluating procedures that help provide reliable county estimates in an efficient manner in their State. The automony in each program is primarily a function of the funding received from the different State cooperators. The NASS County Estimates System provides a common framework for producing county estimates within each State. However, the actual sampling and estimation methods still vary to some degree. Some documented research has been conducted over the years to evaluate different procedures. But the Census of Agriculture continues to be the major evaluation tool.

Ford, Bond, and Carter (1983) examined a model-based approach that estimates the percentages of the total USDA State level crop acreage allocated to each county and district. A composite estimator was used to estimate North Carolina county and district level percentages for 1981. The composite included the estimated percentages based on direct estimates of crop acreage from two separate probability crop acreage surveys and the estimated percentage from a simple linear regression on the percentages over time (1972-1980). The time trend component tended to have much larger weights than the survey components in the composite. Results demonstrated that indications from this procedure were more stable and closer to published values than indications from either of the separate crop acreage surveys. Since the published values tended to follow the composite which is strongly influenced by the time trend model, the results suggested that NASS statisticians were already informally following the linear time trends in setting the county estimates, and consequently, these procedures were never implemented.

The major evaluation process of the NASS county estimates continues to be the review against the Census of Agriculture numbers every five years. NASS statisticians are actually involved in the review of the Census numbers before they are published to resolve any major discrepancies based on their knowledge of the State's agriculture and their county estimates for the comparable year. After this review, the Census data are resummarized and published. NASS State offices then go through the "Census Review" process. The county estimates series during the last five years is reviewed for consistency with the Census numbers and any necessary changes are made. This is a subjective process, and handled differently in each State. Other available check data may also be used in the revision process, such as data from livestock or crop associations.

7.5 Current Problems and Activities

NASS is conducting research on general small area estimation methodology through a cooperative agreement with the Department of Statistics, The Ohio State University. In addition, research needs are being identified by the developers and users of the County Estimates System as they gain experience with the programs.

The methodology research with The Ohio State University has focused on statistical procedures for non-probability survey data with the constraint that the sum of the county estimates must sum to the official NASS State estimate. Initial research considered a multiple regression estimator for obtaining county estimates of wheat production in Kansas (Stasny, Goel, and Rumsey 1991). The regression model is of the form:

$$Y_{ci} = \beta_o + \beta_1 X_{ci1} + \dots + \beta_j X_{cij} + \epsilon_{ci}$$

where:

Y_{ci} = value of dependent variable for the ith record, county c
X_{cij} = value of jth independent variable for the ith record, county c
β_j = j^{th} regression parameter.

Fitted model parameters are obtained from the survey data set of individual farm records. The county total for the cth county may then be estimated as:

$$\hat{T}_{(REG),c} = \sum_{i=1}^{N_c} \left[\hat{\beta}_o + \hat{\beta}_1 x_{ci1} + \hat{\beta}_2 x_{ci2} + \dots + \hat{\beta}_j x_{cij} \right]$$
$$= \hat{\beta}_o N_c + \hat{\beta}_1 x_{c.1} + \hat{\beta}_2 x_{c.2} + \dots + \hat{\beta}_j x_{c.j}$$

where:

N_c =   number of farms, county c

$x_{c,j}$ =   total of the jth independent variable, county c.

The county total can be estimated if county level values are known for all independent variables in the regression model. In the initial analysis of wheat production county estimates, the independent variables were planted acres of wheat and a district indicator which accounted for differences in yield for different areas of the State. Since production is closely related to planted acres and yield, these seem to be reasonable independent variables. It may be more difficult to identify independent variables for estimated planted acreage. These indications would then be scaled by some method. Evaluation of the regression estimator using simulated data indicated that it generally produced more precise indications than a direct expansion of sample data within the respective county. Analysis also indicated that a constant proportional scaling method worked just as well as more sophisticated methods involving the sum of squared differences or the sum of squared relative differences between the county indications and the final estimates. Future research is planned to consider other variables and other small-area estimators.

Research is also being conducted through the cooperative agreement with The Ohio State University on a synthetic estimator for counties that have zero or only a few positive records for a commodity. In spite of the improved sampling capabilities of the new system, this situation still occurs. Approaches that share information from neighboring counties and across States are being investigated.

Also, there is a need to evaluate survey estimates (direct expansion and ratio) generated on a probability basis. The current program combines data from different sampling designs in such a manner that the actual selection probabilities are not used. This procedure was chosen because it is easy to implement. Also, it makes use of all data collected. As stated previously, the same questionnaire is used for all sample units, regardless of the original sampling design. Consequently, barley data are collected from the barley design, from the corn design, from the hog design, etc.. An alternative approach that also makes use of all data collected is to first generate, for each commodity, probability based estimates independently from each design. That is, generate separate barley acreage estimates from the barley design, from the corn design, from the hog design, etc., using the appropriate selection probabilities. These estimates can then be combined to produce an unbiased (or nearly unbiased) estimator with less variance than an estimate based on a single design. Analysis is being conducted to evaluate alternative post-stratification and composite estimation strategies.

As has been described, the NASS County Estimates Program has evolved over the past 70 years. The published estimates continue to be a relied upon source of essential information for many data users in the agricultural community. However, there is a constant concern about the quality of the estimates and methodological improvements that could be made. The program requires a major commitment of resources for the editing, summarization, and publishing of the data. These issues will continue to be a focus of future research as resources allow.

REFERENCES

Bass, J., Guinn, B., Klugh, B., Ruckman, C., Thorson, J., and Waldrop, J. (1989), "Report of the Task Group on County Estimates," National Agricultural Statistics Service, U.S. Department of Agriculture.

Brooks, E. M. (1977), "As We Recall: The Growth of Agricultural Estimates, 1933-1961," Statistical Reporting Service, U.S. Department of Agriculture.

Ford, B. L., Bond, D., and Carter, N. (1983), "Combining Historical and Current Data to Make District and County Estimates for North Carolina," Staff Report AGES 830906, Statistical Reporting Service, U.S. Department of Agriculture.

North Carolina Agricultural Statistics Service (1986), "North Carolina Probability A&P and County Estimates Surveys," Raleigh, NC: Author.

Stasny, E. A., Goel, P. K., and Rumsey, D. J. (1991), "County Estimates of Wheat Production," Survey Methodology, Vol. 17, pp 211-225.

U.S. Department of Agriculture (1917), "Conference of Agricultural Statisticians," Author.

U.S. Department of Agriculture, Bureau of Agricultural Economics (1933), "The Crop and Livestock Reporting Service of the United States," Misc. Publication No. 171, Author.

U.S. Department of Agriculture, Bureau of Agricultural Economics (1949), "The Agricultural Estimating and Reporting Services of the United States Department of Agriculture," Misc. Publication No. 703, Author.

U.S. Department of Agriculture, Agricultural Marketing Service (1957), "National Conference of Agricultural Statisticians: Conference Papers, Part B, Commodity Branch Sessions," Author.

U.S. Department of Agriculture, Statistical Reporting Service (1969), "The Story of U.S. Agricultural Estimates," Misc. Publication No. 1088, Author.

U.S. Department of Agriculture, National Agricultural Statistics Service (1989), "Standard for Suppressing Data Due to Confidentiality," Policy and Standards Memorandum No. 12-89, Author.

U.S. Department of Agriculture, National Agricultural Statistics Service (1991), "Sampling Frame Standards for Coverage and Maintenance," Policy and Standards Memorandum No. 14-91, Author.

U.S. Department of Agriculture, National Agricultural Statistics Service (1992), "Estimation Manual," Volume 10, Author.

CHAPTER 8

Model Based State Estimates from the National Health Interview Survey

Donald Malec
National Center for Health Statistics

8.1 Introduction and Program History

There is a continuing need to assess health status, health practices and health resources at both the national level and subnational levels. Estimates of these health items help determine the demand for quality health care and the access individuals have to it. Although National Center for Health Statistics (NCHS) survey data systems can provide much of this information at the national level, little can be provided directly at the subnational level, except for a few large states and metropolitan areas. The need for State and substate health statistics exists, however, because health and health care characteristics are known to vary geographically. Also, health care planning often takes place at the state and county level.

In this chapter our focus will be the production of state and substate indirect estimators from the National Health Interview Survey (NHIS). Information on health status, health practices and health resources is collected annually in the NHIS and direct national estimates of these items are also produced annually. The NHIS is a multistage, personal interview sample survey. It is redesigned every ten years, in order to make use of new population data collected in the U.S. Census of Population. The sample design fielded in 1985-94 uses 1,983 primary sampling units (PSU's), each PSU consisting of a single county or a group of contiguous counties (minor civil divisions are used instead of counties in New England and Hawaii). The population of 1,983 PSU's is stratified and approximately 200 are sampled with probability roughly proportional to their population sizes. Within each sampled PSU clusters of households are formed and sampled. Areas within a PSU with a high concentration of Blacks are oversampled. The NHIS is a cross-sectional survey, each year a new sample containing approximately 50,000 households and 120,000 individuals is selected. For additional details about the design of the NHIS see Massey *et al.* (1989).

High costs are the primary reason that NCHS is unable to provide subnational estimates from its national surveys. With the current budget, the sample size in most states is often too small to produce precise direct estimates. There is also a concern that direct estimates of small areas will have a larger component of nonsampling error. For example, a small area may be only canvassed by one interviewer and the resulting direct estimate may be affected by this interviewer's style. In contrast, direct estimates that cover higher geographic levels are canvassed by many interviewers and will tend to have a smaller interviewer affect due to the involvement of many independent interviewers. However, even with large sample sizes and many interviewers, problems can occur in preparing direct estimates of the variance of direct state estimates because the NHIS was not designed for this purpose (Parsons, Botman and Malec, 1990).

Since 1968, the National Center for Health Statistics has produced and evaluated indirect state estimators of health items derived from the NHIS. Although NCHS does not have a program for the regular publication of subnational estimates based on indirect estimation methods, it has supported the development and evaluation of these techniques. This aim has been achieved through the support of in-house research, research grants and small-area conferences and workshops. Through these efforts, three Public Health Series reports, containing indirect State estimates of disability and the use of health care have been published (see section 8.2.1). A number of methodological research projects have also taken place at the Center. Research results have appeared in the Center's series reports and in journals and conference proceedings. The demand for small area estimates is increasing. Subnational estimates are sometimes needed for the administration of Federal Block grants. States are also striving to meet the health guidelines for the year 2000 as promoted in *Healthy People 2000: National Health Promotion and Disease Prevention Objectives*. The assessment of the dietary and nutritional status of the U.S. requires an understanding of these factors at the subnational level. Accurate estimates are needed for all these purposes.

The Center is continuing research efforts into the development of subnational estimators. Currently, estimates based on a hierarchical, logistic regression are being produced and evaluated. This model includes demographic effects and county level effects and includes county level variation. These continuing efforts are being made to both improve the accuracy of small area estimates and to produce estimates of their accuracy.

The current NHIS, which will be in the field from 1995 to 2004, will oversample and screen for both Blacks and Hispanics. Approximately twice as many PSUs will be sampled as were sampled in the 1985-94 design. In addition PSUs will be stratified by state and by urban/rural PSUs within a state. This stratification will not produce precise state estimates but it will provide a convenient framework for supplementing the NHIS state data with additional state data. The use of state strata may also improve indirect state estimates.

8.2 Program Description, Policies and Practices

8.2.1 Estimates from the NHIS

Three reports containing indirect state estimates from the NHIS have been published.

The first report, *Synthetic State Estimates of Disability Derived from the National Health Survey* (NCHS, 1968), contains estimates of long- and short-term disability measures collected during July 1962 - June 1964. Specifically, the report contains the percent of persons who suffer from one or more chronic conditions, the percent of persons whose activity is limited by a chronic condition, the average number of restricted activity days per person, the average number of bed disability days per person and the average number of work-loss days per employed person. Estimates were made using a ratio adjusted synthetic estimate.

The second report, *State Estimates of Disability and Utilization of Medical Services: United States, 1969-1971* (NCHS, 1977), also contains state estimates of disability as well as state estimates of short-stay hospital utilization, physician visits and dental visits. The estimates in this report are also ratio adjusted synthetic estimates.

The third report, *State Estimates of Disability and Utilization of Medical Services: United States, 1974-76* (NCHS, 1978) contains estimates of the same health items as the preceding report; disability, short-stay hospital utilization, physician visits and dental visits. These estimates were made using a composite estimation method.

These reports present estimates of levels but contain no estimates of accuracy. The reason estimates of accuracy were not presented is because satisfactory estimates of the error of individual estimates of states did not exist for either synthetic or composite estimates.

8.2.2 Small Area Research Conferences

The Center has sponsored or cosponsored three research conferences on small area estimation. The first conference, cosponsored with the National Institute on Drug Abuse was held in Princeton, N.J. in 1978. The second conference was held in Snowbird, Utah, in 1984. The third conference was held in New Orleans, LA, in 1988. The first two conferences produced published proceedings (see NIDA Research Monograph 24, 1979, and NCHS, 1984).

8.3 Estimator Documentation

NCHS has no regular program of producing indirect state estimates from the NHIS. However, the following estimators have been used in the past to prepare state estimates of health characteristics. Many of these estimators were introduced to correct for the known deficiencies of the synthetic estimator.

8.3.1 Basic Synthetic Estimator

The basic synthetic estimator is used for State estimation when national estimates by class and State-specific population counts by class are available. The synthetic estimator weights the national class means by the proportion of persons in the state belonging to the class.

The form of the estimator for state d is:

$$\bar{y}_{(SYN),d} = \sum_{b=1}^{B} p_{bd}\bar{y}_{(E),b}$$

where $\bar{y}_{(E),b}$ is the national level, direct, design-based expansion estimator of the average health characteristic of individuals in class b. The value of p_{bd} is the proportion of civilian non-institutionalized individuals in state "d" belonging to demographic class "b".

A synthetic estimator is unbiased if the population can be divided into mutually exclusive and exhaustive classes, b, in which the average health characteristic in each class does not vary among small areas. If this assumption is true and if a large enough sample is selected in each class, then the synthetic estimator will be accurate.

In chapter two, entitled "Synthetic Estimation in Followback Surveys at the National Center for Health Statistics", a detailed example of the use of synthetic estimation is provided. Another example of the construction of a synthetic estimate can be found in Schaible *et al.* (1977) where they create sixty-four demographic classes based on cross-classifications of variables defined by race, sex, age, family size and industry occupation of head of family.

The values of p_{bd} based on this cross-classification were obtained from the U.S. Census. Specifically, mid-decade estimates of the state civilian population were ratio adjusted by population census counts to produce estimates for the sixty-four demographic classes for the civilian, non-institutionalized population. For each of the sixty-four U.S. subpopulations, the NHIS was used to obtain the estimates, $\bar{y}_{(E),b}$. By combining these components together, as detailed in the above formula, a synthetic estimate can be constructed for each state.

In addition, the synthetic estimator, $\bar{y}_{(SYN),d}$, can be used to produce synthetic estimates for proportions and totals. Since a proportion is the average of measurements coded as zero or one, the synthetic estimator for a proportion is identical to the synthetic estimator of a mean. A synthetic estimate for a total can be obtained by multiplying $\bar{y}_{(SYN),d}$ by the total population of area d. That is:

$$\hat{Y}_{(SYN),d} = N_d \bar{y}_{(SYN),d}$$

where N_d is the population of state d.

The synthetic estimator appears in a number of NCHS related publications (e.g., NCHS, 1968, 1977, Levy, 1971 and Namekata, Levy and O'Rourke, 1975) and has been used extensively.

8.3.2 Ratio Adjusted Synthetic Estimator

When regional, direct estimates are available, state synthetic estimates are often ratio adjusted to their regional direct estimate. In this way regional estimates, obtained by combining synthetic State estimates in a region together, will equal the corresponding direct estimator. This adjustment removes all bias in the synthetic estimate at the regional level and is an attempt to remove bias of the synthetic estimator at the state level.

The form of this estimator is:

$$\bar{y}_{(SYNR),d} = a\bar{y}_{(SYN),d}$$

where $\bar{y}_{(SYN),d}$ is the basic synthetic estimator and $a = \bar{y}_{(E),R}/\sum_{d \in R} p_d \bar{y}_{(SYN),d}$ is the ratio of a design based estimate of $\bar{Y}$ for region R and the corresponding synthetic estimate. This regional synthetic estimate is obtained by weighting the synthetic estimate of the state's average by its proportion of civilian noninstitutional population in the region, p_d, and then summing over all states in region R.

This estimator has also been used in a number of NCHS publications (e.g., NCHS, 1968, 1977, and Levy, 1971). It is also used by the Department of Agriculture in their NASS county estimates program (see chapter 7).

8.3.3 Composite Estimation

In small area estimation, an unbiased, direct estimate of the area can usually be constructed. However, it is often not precise enough to be of use, by itself. However, it can be used to reduce the bias (and, ultimately, the Mean Square Error) of an indirect estimator. This is done by forming a composite estimator which is a weighted average of the direct and indirect estimators. In general, a composite estimator combines two estimators together resulting in an estimator which may be more accurate than either of its components. A composite estimator of $\bar{Y}$ consists of taking two, possibly biased, estimators, $\bar{y}_1$ and $\bar{y}_2$ and forming a weighted average, $\bar{y}_c = \lambda \bar{y}_1 + (1-\lambda) \bar{y}_2$. The ideal value of λ is selected to minimize the Mean Square Error (MSE) of $\bar{y}_c$. That is, λ is chosen to minimize:

$$\mathrm{MSE}(\bar{y}_c - \bar{Y}) = \lambda^2 \mathrm{MSE}(\bar{y}_1) + (1-\lambda)^2 \mathrm{MSE}(\bar{y}_2)$$

$$+ 2\lambda(1-\lambda)(\mathrm{Cov}(\bar{y}_1, \bar{y}_2) + (\bar{Y} - E(\bar{y}_1))(\bar{Y} - E(\bar{y}_2))$$

If one of the component estimators is unbiased (as described above) and if $\mathrm{Cov}(\bar{y}_1, \bar{y}_2)$ is small relative to both $\mathrm{MSE}(\bar{y}_1)$ and $\mathrm{MSE}(\bar{y}_2)$ then the optimum value of λ is approximately (Schaible, 1979):

$$\lambda \doteq \frac{\mathrm{MSE}(\bar{y}_2)}{\mathrm{MSE}(\bar{y}_1) + \mathrm{MSE}(\bar{y}_2)} = \frac{1}{1 + R}$$

where $R = \mathrm{MSE}(\bar{y}_1)/\mathrm{MSE}(\bar{y}_2)$.

Although any two estimates can be combined together to form a composite estimate, a ratio-adjusted synthetic estimate combined with a direct estimate has been used to publish state estimates (NCHS, 1978).

The form of this estimator is:

$$\bar{y}_{(COMP),d} = w_d \, \bar{y}_{(E),d} + (1 - w_d) \, \bar{y}_{(SYNR),d}$$

where $\bar{y}_{(E),d}$ is an unbiased direct estimate of $\bar{Y}_d$ and $\bar{y}_{(SYNR),d}$ is a ratio-adjusted synthetic estimate of the same characteristic. The two component estimators are weighted by $w_d = 1/(1 + \hat{R}_d)$, where the quantity, $\hat{R}_d$, is an estimate of the ratio of the mean square error of $\bar{y}_{(E),d}$ to the mean square error of $\bar{y}_{(SYNR),d}$. The composite estimator can perform well even if a relatively inaccurate estimate of this ratio is used (Schaible, 1979). As a result, model based techniques have been used to estimate $\hat{R}_d$ (Schaible, 1979).

References and uses of composite estimators can be found in (NCHS, 1968, 1978), Schaible *et al.* (1977) and Schaible (1979).

A number of other estimators have been developed for evaluations but have not been used to produce estimates in NCHS publications.

8.3.4 Regression-Adjusted Estimator

Recognizing that a synthetic estimate of a small area may not account for local influences, the regression-adjusted estimator does so by incorporating covariates, available at the local level, with a synthetic estimate. The covariates are used to model the bias associated with the synthetic estimate. The model is then used to produce a regression-adjusted estimator, which incorporates an adjustment of the bias. The method as applied in Levy (1971) and Levy and French (1977) begins with state level synthetic estimates but models the bias at the regional level, instead of at the state level. After fitting a linear model to the relative biases at the regional level, the estimated parameters in the model are used to adjust the synthetic estimates at the state level.

The form of the estimator is:

$$\bar{y}_{(RA),d} = \bar{y}_{(SYNR),d}\left(1 + a + bX_d\right)$$

where $\bar{y}_{(SYNR),d}$ is the ratio adjusted synthetic estimator of state d and X_d is a covariate measured on state d. The values of a and b are the least squares estimates of α and β, obtained from the following model at the region level:

$$\bar{y}_{(E),R} = \bar{y}_{(SYNR),R}\left(1 + \alpha + \beta X_r\right) + \text{error}.$$

There are other ways to use covariates with a synthetic estimator, such as incorporating them directly into the estimator. This approach has been used by Elston, Koch and Weissert (1991) in the modeling and production of disability status using county covariates. Based on a loglinear model of discrete states of disability characterized by both demographic class indicators and county covariates, estimates for States and large counties have been made. See Elston, Koch and Weissert for more details.

8.3.5 Nearly Unbiased Estimator

The nearly unbiased estimator is a special case of the basic synthetic estimator where the "demographic classes", b, are defined to be the sampling strata. This estimator takes the usual direct estimates of the strata means and allocates them to a particular state based on the proportion of the state population in each stratum. The form of the estimator for state d is:

$$\bar{y}_{(NU),d} = \sum_{b=1}^{B} P_{bd}\, \bar{y}_{(E),b}$$

where $\bar{y}_{(E),b}$ is the national level, direct, design-based expansion estimator of the average health characteristic of individuals in stratum b. The value of p_{bd} is the proportion of civilian non-institutionalized individuals in state "d" belonging to stratum "b". When the composition within each stratum is relatively homogeneous, this estimator will be "nearly unbiased". See NCHS (1968) and Levy and French (1977) for more details.

8.3.6 Woodruff's Estimator

Based on a regression estimator constructed at a higher level of geography, Woodruff's method consists of applying the regression coefficient, estimated at the higher level of geography, to a corresponding regression estimator at the small area. In simplified form, the estimator is:

$$\bar{y}_{(W),d} = \bar{y}_{(E),d} + B(\bar{X}_d - \bar{x}_{(E),d})$$

where $\bar{y}_{(E),d}$ and $\bar{x}_{(E),d}$ are the direct estimators of the mean of y and x, respectively, for the small area, d. The population average of x for the small area d is denoted by $\bar{X}_d$ and the estimated regression coefficient, estimated at the higher level of geography, is denoted by B.

This estimator is an example of a generalized difference estimator (see Särndal, 1984, for an account of generalized difference estimators used in small area estimation). For more details, see Woodruff (1966). For a comparison with synthetic estimates, see NCHS (1968).

8.3.7 Bayesian Methods Utilizing Hierarchical Models: General

A hierarchical model is currently being used to model the geographic variation of binary health variables (e.g., the presence or absence of a particular health practice). The major aim is to account for the small area variation that is ignored by synthetic estimates. As part of this goal, county level covariates are being used in the modelling to improve the predictability for small areas. Additionally, the variability of responses due to local effects is explicitly incorporated into the model. Estimates and their accuracy are made using Bayesian predictive inference. The production of state estimates, using this methodology is being planned.

This method is being used to predict binary outcomes, for example the percentage of persons in a state who have visited a physician in the past year. Let Y indicate whether an individual has $(Y=1)$ or has not $(Y=0)$ visited a physician in the past year. Let p_{cb} denote the probability that an individual in demographic class b and county c visits a physician in the past year. Then, the chance that individual, k, in demographic class, b, in county, c, visits a physician in the past year is denoted as:

$$Pr(Y_{cbk} = y_{cbk}) = p_{cb}^{y_{cbk}}(1-p_{cb})^{1-y_{cbk}}, \qquad y_{cbk} \in \{0,1\}. \tag{1}$$

In addition, assume that each observation is independent within a county.

Further, define

$$\ln(p_{cb}/1-p_{cb}) = X_{cb}\boldsymbol{\beta}_c, \tag{2}$$

where X_{cb} is a vector of covariates which characterizes the demographic class, b. This model defines a logistic regression within county c.

The county logistic parameter is allowed to vary from county to county, thereby incorporating local error into the model. Specifically,

$$\boldsymbol{\beta}_c \sim N(Z_c\boldsymbol{\eta}, \boldsymbol{\Gamma}), \tag{3}$$

where Z_c is a matrix of county covariates (e.g. the percent of adults in a county who have completed high school or county per-capita income). Each row of Z_c corresponds to a linear relationship to the corresponding element of $\boldsymbol{\beta}_c$. Finally, a distribution expressing little prior knowledge about η and Γ is specified; i.e., $p(\eta,\Gamma)$ ∝ constant.

The objective is to estimate the average for state, d:

$$\bar{Y}_d = \sum_{c\in d}\sum_b\sum_k Y_{cbk}/N$$

by predicting the values of Y_{cbk} which have not been sampled.

The posterior mean of $\bar{Y}_d$ is used as an estimate of $\bar{Y}_d$. It takes the form of:

$$\bar{y}_{(H),d} = \sum_{c\in d}\sum_{(b,k)\in s} Y_{cbk}/N + \sum_{c\in d}\sum_b \{N_{cb} - n_{cb}\} E(p_{cb}|y_s)/N$$

where $E(p_{cb}|y_s)$ denotes the posterior mean of p_{cb}. Here, N_{cb} and n_{cb} are, respectively, the population and sample sizes in class b of county c. Note that N_{cb}, the total number of persons in county c and demographic group b, is needed to produce this estimate. Typically, estimates are available when b corresponds to a cross-classification of age by race by sex.

The form of the model (i.e. the choice of X_b and Z_c is determined using data analysis tools, such as scatterplots, partial logistic residual plots and stepwise logistic regression. Unfortunately, $\bar{y}_{(H),d}$ (and the posterior variance) do not have a closed form expression and must be evaluated numerically. This is accomplished using Markov Chain Monte Carlo methods, specifically the Gibbs sampler (see, e.g., Smith and Roberts, 1993).

The hierarchical model, described in (1), (2) and (3), is equivalent to a nonlinear random effects model. From (2) and (3), one has that: $\beta_c = Z_c\eta + \varepsilon_c$, where $\varepsilon_c \sim N(0,\Gamma)$, so that the logistic term can be rewritten as: $\ln(p_{cb}/1-p_{cb}) = X_{cb}Z_c\eta + X_{cb}\varepsilon_c$. In this form, the model is a generalization of the mixed model used by MacGibbon and Tomberlin (1989) to construct small area estimates. Instead of including only one random effect term for each county (as MacGibbon and Tomberlin do), the interaction of random county effects with individual level covariates are included in this generalization. At the other extreme, if Γ is set to zero, indicating no random effects, then a model corresponding to a type of regression-adjusted synthetic estimator results. The hierarchical model, described here, corresponds to the model used by Wong and Mason (1985). However, Wong and Mason used an empirical Bayes approach, treating the estimate of Γ as fixed. Utilizing the Gibbs sampler approach, all variability due to estimation is accounted for. For a description of the distinction between empirical Bayes methods and Bayesian Hierarchical analysis see Berger (1985), sections 4.5 and 4.6.

Approximate methods of inference that account for the fact that η and Γ are unknown are available in both a Bayesian and frequentist framework. These approximate methods may be less computationally intensive than the Gibbs sampler approach. Kass and Steffey (1989) present a general, approximate method in a Bayesian framework. When the model is a Normal, using a frequentist approach, Prasad and Rao (1990) determine an approximate mean square error for estimators and apply the results to specific indirect, small area estimators.

8.3.8 Example of Bayesian Methods Utilizing Hierarchical Models: Some Health Care Results

The following summarizes preliminary work on the modeling and production of small area estimates of the percent visiting a physician in the past year, as measured in the NHIS. For a more detailed account see Malec, Sedransk and Tompkins (1993) and Malec and Sedransk (1993). The summary is presented in three parts; i) Selection of variables for the model, ii) estimation of $\overline{Y}_d$ and iii) comparison of the estimates with a synthetic estimate and a direct estimate.

i) Selection of Variables: Given the basic model, specified in (1), (2) and (3) of section 8.3.7, the first step of the method entails selecting the variables to use in the model (i.e., the specification of X_b and Z_c). To avoid an inordinate number of possible variables for inclusion, individual level variables (i.e., candidates for X_b) were selected first, then county covariates (i.e., candidates for Z_c) were selected second. To accomplish this, the random variation between counties was ignored (i.e., Γ, the variability of β_i is assumed to have little effect on the selection of variables and was set at zero).

In order to evaluate whether a specific demographic trait belongs in the model, partial residual plots were used. After fitting a simple age model: $\ln(p_{cb}/1-p_{cb}) = \beta_1 + \text{age}_b \cdot \beta_2$, the residuals, $r_{cbk} = (y_{cbk} - \hat{p}_{cbk})/\hat{p}_{cbk}(1-\hat{p}_{cbk})$, were determined and averaged within age by sex groups (see Fienberg and Gong's comment to Landwehr, Pregibon and Shoemaker 1984). Figure 1 shows that these partial residuals still exhibit age and sex effects.

Several key features can be seen in this figure. There is a predominant sex effect. While relatively fewer males in their 20's or 30's visit a physician, relatively more females visit a physician during these child-bearing years. In addition, after accounting for a linear age term, the relative propensity to visit a physician increases for both the underaged and the overaged, regardless of sex.

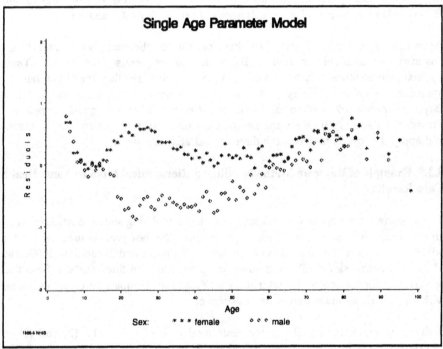

Figure 1: Average residuals of percent visiting a physician in past year. Based on a single parameter age model.

To account for these effects, independent variables corresponding to linear splines were used. After examining a number of residual plots by age, race and sex, a final set of independent variables was chosen. Based on visual inspection, race effects were considered negligible and not included in the final model. In relation to an individual in county, c, in the age and sex class denoted by, b, the final set of independent variables are defined as follows:

$$X_{cb1} = 1$$

$$X_{cb2} = \text{age}_b$$

$$X_{cb3} = I_{[\text{male}]} \times (\text{age}_b - 13)_+$$

$$X_{cb4} = I_{[\text{female}]} \times (\text{age}_b - 13)_+$$

$$X_{cb5} = I_{[\text{female}]} \times (\text{age}_b - 25)_+$$

$$X_{cb6} = I_{[\text{male}]} \times (\text{age}_b - 25)_+ + I_{[\text{female}]} \times (\text{age}_b - 35)_+$$

$$X_{cb7} = (\text{age}_b - 65)_+$$
$$X_{cb8} = (\text{age}_b - 79)_+,$$

where $(x)_+$ represents the positive part of x, age_b is the age associated with demographic group, b, and $I_{[\text{sex}]}$ is 1 if sex matches the sex associated with demographic group, b, and 0 otherwise. The eight variables represent an intercept term, a linear age effect and linear age splines with sex interactions. In particular, X_{cb6} represents separate, but parallel, linear age splines for males and females.

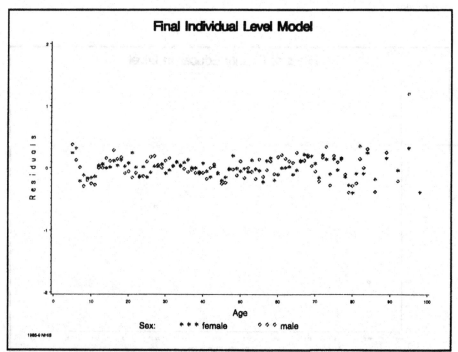

Figure 2: Average residuals for the percent visiting a physician in the past year. Based on final individual level model.

Partial residual plots based on this eight parameter model were computed and figure 2 plots the averages of these residuals within age by sex group. The partial residual plot indicates a relative absence of age and sex affects. (The apparent heterogeneity in the plots is at least partly due to an unequal sample size in each age group.)

These residuals, based on the eight parameter model outlined above, are then used to assess the affect of county covariates. The resulting residuals, with the individual age and sex effects removed, were then averaged within counties of a given type, for example counties with a high level of educational attainment. Corresponding to various typologies of counties, plots of the residuals versus county types were used to assess the influence of county covariates on the proportion of persons visiting a physician. Figure 3 illustrates one such comparison. Here, residuals are averaged within counties exhibiting a certain education level. A number of independent variables were examined in this manner. For physician visits, economic type variables such as per capita income, percent of population below poverty and education level exhibited similar trends. Other county covariates from the March 1989 Area Resource File (1989) and the NCHS County Mortality files were also examined.

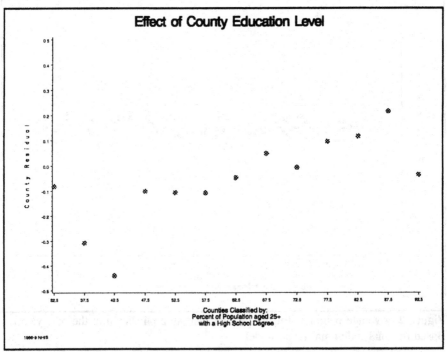

Figure 3: Residuals based on final individual level model averaged over counties grouped by the percent of persons, aged 25+, with a high school degree.

Based on the examination of the residuals, the following specific county level model was used for the specification in (3):

$$E(\beta_{ch}|\eta) = \eta_{h1} + EDHS_c\eta_{h2}$$ (5)

Where, $h = 1, ..., 8$, and $EDHS_c$ is the percent of the population aged 25+ in county c with at least a high school degree. The components of the vector, $\eta = (\eta_{11}, \eta_{12}, \eta_{21}, ..., \eta_{81}, \eta_{82})$, account for a separate intercept and slope term for each of the eight individual effects.

ii) Estimation: Based on the model introduced and the variables selected, the posterior mean, $\bar{y}_{(H),d}$, and the posterior variance can be evaluated numerically using the Gibbs sampler (see Malec, Sedransk and Tompkins 1993). By combining four years, 1986-1989, of NHIS responses together, a normal approximation to the likelihood of β_i was used. Using this approximation, the conditional posterior distributions needed to implement the Gibbs sampler, can be derived in a straightforward manner.

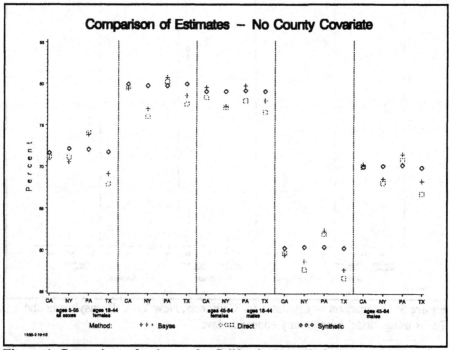

Figure 4: Comparison of estimates for California, New York, Pennsylvania and Texas using model with no county covariates.

iii) Comparison of Estimates: Estimates were produced in the manner described above and, for large states, compared to direct estimates. For large states, the direct estimates will be relatively precise. In addition, for large states, the borrowing of local data should show up in the Bayes estimates. For contrast, a type of synthetic estimate was also used in the comparison. The synthetic estimate is developed using

the same model as specified by (1) through (5) but assuming $\Gamma = 0$. The Maximum
Likelihood Estimates (MLEs) of η were then obtained. The synthetic estimate was
constructed using these MLEs. Figures 4 and 5 show comparisons for four large
states for the three types of estimates both without and with the county covariate,
EDHS. The Bayes estimate is generally closer to the direct estimate and exhibits
more variation than the synthetic estimate. In figure 5, the use of the county
covariate increases the variability of the synthetic estimates from state to state but,
often, not in the same direction as the direct state estimate or Bayes estimate.

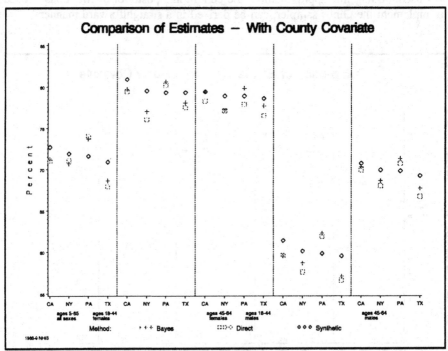

Figure 5: Comparison of estimates for California, New York, Pennsylvania and
Texas using model with county education level.

Based on this preliminary work, State estimates for both physician visits and health
status are being planned for publication.

8.4 Evaluation Practices

8.4.1 Reports of Evaluation Studies

A number of evaluations on small area estimators have been conducted at NCHS. The first publication of synthetic estimates (NCHS, 1968) includes a comparison of a synthetic estimator, a nearly unbiased estimator and Woodruff's estimator. Since then a number of other evaluation studies have been conducted at the Center. A short description of each is given below.

The Use of Mortality Data in Evaluating Synthetic Estimates

A basic synthetic estimator, a ratio adjusted synthetic estimator and a regression adjusted synthetic estimator are evaluated by constructing these estimates of mortality rates for Motor Vehicle Accidents, Major Cardiovascular Diseases, Suicides and Respiratory T.B. using the complete population of mortality events compiled by NCHS. These state estimates are then compared to their known rates, using the same data. It was found that the accuracy of the synthetic estimates varied considerably from state to state and from item to item. (Levy, 1971).

Synthetic Estimates of Work Loss Disability for Each State and the District of Columbia

The basic synthetic estimates of partial and complete work disability are compared to precise estimates from the 1970 Census of Population and Housing. Here, the agreement between synthetic and direct estimates of partial work disability were found to be fairly good while the agreement between corresponding estimates of complete work disability were fairly poor. (Namekata, Levy and O'Rourke, 1975)

Synthetic Estimates of State Health Characteristics Based on the Health Interview Survey

Formulas for the bias and variance of the nearly unbiased estimator and the ratio-adjusted synthetic estimator are developed. Correlations and average percentage errors between state synthetic estimates based on different formulations of demographic groups are determined for each of a number of NHIS items. For the cases evaluated here, the coarseness of the demographic groupings had little effect on the resulting synthetic estimates. (Levy and French, 1977)

An Empirical Comparison of the Simple Inflation, Synthetic and Composite Estimators for Small Area Statistics

Using NHIS data, state unemployment rates and percent completing college were estimated using a simple direct estimator, a synthetic estimator and a composite estimator. Each of these estimates were compared to accurate estimates obtained from the 1970 Census. It was demonstrated that the composite estimator was much more accurate than either the synthetic estimator or the inflation estimator for the items under study. (Schaible, Brock, Casady and Schnack, 1977)

8.4.2 Comparison with a known standard

The following is a summary of the types of evaluation methods that have been used in the aforementioned reports.

Measurements such as: work-loss disability, unemployment rates, percent completing college and marital status are compiled in the U.S. Population Census and are known exactly or with a high degree of accuracy for small geographic areas. In addition, vital statistics such as the mortality rates from motor vehicle accidents, Cardiovascular disease, suicides and Tuberculosis are known exactly for counties. Although these measurements are not the same as the morbidity rates or health care utilization rates, measured in the Center's surveys, they are related. Small area estimation procedures have been tested by making state estimates for these quantities and comparing them to the known quantity. This technique has shown that synthetic state estimates will often have less variability than their corresponding ensemble of true values (Schaible, Brock, Casady and Schnack, 1977, 1979). By estimating known quantities, some of the deficiencies of an estimation method can be ascertained. However, the fact that a particular method works well in estimating known quantities does not imply that the method will work well in other situations.

When estimating a known quantity, the distribution of the errors is usually presented in summary form. Usually, the average absolute error or the average squared error is calculated. To assess the similarity of two estimation methods, the simple correlation between errors is calculated.

There are other ways to obtain accurate standards for comparison. For example, accurate, direct estimates are often available for the very large states and for groups of states (Levy, 1971). In addition, years of data can be pooled together to create a larger population in which direct estimates can be made for a number of subnational areas and then compared to an indirect estimate (Malec, Sedransk and Tompkins, 1993).

NCHS publications utilizing this method include NCHS (1968); Levy (1971); Malec, Sedransk and Tompkins (1993); Namekata, Levy and O'Rourke (1975); Schaible, Brock, Casady and Schnack (1977, 1979).

8.4.3 Comparison of alternative estimators

An alternative estimator can be constructed to specifically deal with a deficiency in another estimator. For example, a composite estimator of a state will preferentially use its state data, whereas a synthetic estimator will not. The relevance of this problem can be evaluated by comparing the two estimates. This method can be used to improve estimates but, since the true value is not known, the quality of the improved estimate is not known. For example, two types of estimators may yield similar state estimates but both be in error. In contrast, a number of different types of estimators may each yield vastly different estimates but one of them may be accurate.

8.4.4 Model Evaluation with Data Analysis

When a statistical model is used to produce indirect estimates, features of the model can be checked using the data that has been sampled. In fact, data from a national survey can be used to develop a model that fits the data well (Malec, Sedransk and Tompkins, 1993), -- although there may be features of a model that are difficult to evaluate, if the model is complex.

8.5 Current Problems and Activities

The Bayesian method, utilizing a hierarchical model (in section 8.3.8) is currently being refined and evaluated with the aim of producing state estimates based on a single year of data. A relatively efficient method of Gibbs sampling due to Gilks and Wild (1992) has been used to produce state estimates. This procedure uses the exact specifications from (1) and (2) of section 8.3.7 and does not require the pooling of four years of data or a normal approximation to the likelihood. Gilks and Wild's method is still very computationally time consuming and it is being compared to both an alternative normal approximation to the conditional posterior of β_i and to the normal approximation of the likelihood. The comparison is in terms of both accuracy and computational effort.

The variable selection procedure for the hierarchical model has also been improved. Stepwise spline selection is now being used for both individual level variables, county level variables and interactions between individual-level and county-level variables. In addition, more specific methods to evaluate the model fit are being considered. The impact of the design on inference is also being further evaluated.

In addition to the Bayesian hierarchical method presented above, three other research projects have recently been undertaken at NCHS.

- As part of a research contract to evaluate and redesign the NHIS, a generalization of the synthetic estimator is being developed. First, the population is partitioned into mutually exclusive demographic cells, as in synthetic estimation. Then, within each demographic cell, a hierarchical model is specified for the responses among the small areas. A distribution is not specified, only the first two moments are used. Estimators are derived in a Bayesian framework where data-based prior information, also specified by only the first two moments, may be incorporated. Alternative estimators, derived within this framework, are compared. Estimates of mean square error, that are more state specific, are also examined. For more information, see Marker and Waksberg (1993).

- The regression adjusted synthetic estimation method of Elston, Koch and Weissert (1991), (see section 8.3.4 above) has been extended by them, under an NCHS contract, to produce estimates of disability covering individuals of all ages.

- An empirical study utilizing 1990 Census data on disability is also being planned. A subsample of the census data, similar to the NHIS sample, will be used to model and predict states. These estimates will then be compared to values based on the entire census. Both the hierarchical model, in section 8.3.7, and the aforementioned synthetic regression method, of Elston, Koch and Weissert, will be used to produce and compare estimates.

8.6 Summary and Some Conclusions

The National Center for Health Statistics has been developing, producing and evaluating indirect state estimates for over two decades. The newer methods proposed generally incorporate local data with the aim of correcting known deficiencies of established estimates and providing more accurate estimates. Recently, the goal to provide both a good, indirect estimate as well as an estimate of its accuracy has received increased attention. Improvements or deficiencies of indirect estimators have been evaluated using related data for which the actual population values are known. Improvements have also been made by developing classes of estimators which include competing estimators as a special case. Small area estimation research has also followed new developments in computing algorithms and the availability of cheaper computing. In particular, the Gibbs Sampler and related methods have opened up the possibility of utilizing more realistic and complex models for small area estimation. It is safe to say that much more can be accomplished in this domain.

The continued research and evaluation of indirect estimators of small areas has helped educate the consumer of these statistics to be more critical of them and to be aware of the underlying assumptions. The fact that small area estimation research continues to be an active and supported area of research is a testament to the continued demand for detail.

REFERENCES

Berger, J.O. (1985) *Statistical Decision Theory and Bayesian Analysis*, second edition. Springer-Verlag.

Gilks, W.R., and Wild, P. (1992) Adaptive Rejection Sampling for Gibbs Sampling. *Applied Statistics*, 41: 337-348.

Elston, J.M., Koch, G.G., and Weissert, W.G. (1991) Regression-Adjusted Small Area Estimates of Functional Dependency in the Noninstitutionalized American Population Age 65 and Over. *American Journal of Public Health*, 81: 335-343.

Kass, R.E., and Steffey, D. (1989) Approximate Bayesian Inference in Conditionally Independent Hierarchical Models (Parametric Empirical Bayes Models). *Journal of the American Statistical Association*, 84: 717-726.

Landwehr, J.M., Pregibon, D., and Shoemaker, A.C. (1984) Graphical Methods for Assessing Logistic Regression Models. *Journal of the American Statistical Association*, 79: 61-83.

Levy, P.S. (1971) The use of Mortality data in evaluating synthetic estimates. *Proceedings of the American Statistical Association, Social Statistics Section*: 328-331.

Levy, P.S., and French, D.K. (1977) *Synthetic Estimation of State Health Characteristics Based on the Health Interview Survey*. Vital and Health Statistics: Series 2, No. 75, DHEW Publication (PHS) 78-1349. Washington: U.S. Government Printing Office.

MacGibbon, B. and Tomberlin, T.J. (1989) Small Area Estimates of Proportions Via Empirical Bayes Techniques. *Survey Methodology*, 15: 237-252.

Malec D. and Sedransk J. (1993) Bayesian Predictive Inference for Units with Small Sample Sizes: The Case of Binary Random Variables. *Medical Care*, 5: YS66-YS70.

Malec D., Sedransk J. and Tompkins, L. (1993) Bayesian Predictive Inference for Small Areas for Binary Variables in the National Health Interview Survey. In *Case Studies in Bayesian Statistics*, eds., C. Gatsonis, J.S. Hodges, R.E. Kass and N.D. Singpurwalla. Springar-Verlag.

Marker, D.A. and Waksberg, J. (1993) Small Area Estimation for the U.S. National Health Interview Survey. In *Small Area Statistics and Survey Designs*, Vol. 1, Central Statistical Office, Warsaw, Poland.

Massey, J.T., Moore, T.F., Parsons V.L. and Tadros, W. (1989), "Design and Estimation for the National Health Interview Survey, 1985-94," National Center for Health Statistics. Vital and Health Statistics, 2:110.

Namekata, T., Levy, P.S., and O'Rourke, T.W. (1975) Synthetic Estimates of Work Loss Disability for Each State and the Districts of Columbia. *Public Health Reports*, 90: 532-538.

National Center for Health Statistics. (1968) *Synthetic State Estimates of Disability*. PHS Publication No. 1759. U.S. Government Printing Office.

National Center for Health Statistics. (1977) *State Estimates of Disability and Utilization of Medical Services, United States, 1969-1971*. DHEW Publication No. (HRA) 77-1241. Health Resources Administration. Washington: U.S. Government Printing Office.

National Center for Health Statistics. (1978) *State Estimates of Disability and Utilization of Medical Services, United States, 1974-1976*. DHEW Publication No. (PHS) 78-1241. Public Health Service. Washington: U.S. Government Printing Office.

National Center for Health Statistics (1984) Invited Papers to the Data Use Conference on Small Area Statistics. *Proceedings of the 1984 NCHS Data Use Conference on Small Area Statistics*, Snowbird, Utah.

NIDA Research Monograph 24 (1979) *Synthetic Estimates for Small Areas* DHEW Publication No. (ADM) 79-801. Health Resources Administration. Washington: U.S. Government Printing Office.

Parsons, V.L., Botman, S.L. and Malec, D. (1990) State Estimates for the NHIS. *1989 Proceedings of the Section on Survey Research Methods, American Statistical Association*. pp. 854-859.

Prasad, N.G.N. and Rao, J.N.K. (1990) The Estimation of the Mean Squared Error of Small-Area Estimators. *Journal of the American Statistical Association*, 85: 163-171.

Särndal, C.E. (1984) Design-Consistent Versus Model-Dependent Estimation for Small Domains. *Journal of the American Statistical Association*, 79: 624-631.

Schaible, W.L., Brock, D.B., Casady, R.J. and Schnack, G.A. (1977) An Empirical Comparison of the Simple Inflation, Synthetic and Composite Estimators for Small Area Statistics. *Proceedings of the American Statistical Association, Social Statistics Section*: 1017-1021.

Schaible, W.L., Brock, D.B., Casady, R.J. and Schnack, G.A. (1979) *Small Area Estimation: An Empirical Comparison of Conventional and Synthetic Estimators for States*. Vital and Health Statistics: Series 2, No. 82, DHEW Publication (PHS) 80-1356. Washington: U.S. Government Printing Office.

Schaible, W.L. (1979) A composite Estimator for Small Area Statistics *Synthetic Estimates for Small Areas* DHEW Publication No. (ADM) 79-801. Health Resources Administration. Washington: U.S. Government Printing Office.

Smith, A.F.M. and Roberts, G.O. (1993) Bayesian Computation Via the Gibbs Sampler and Related Markov Chain Monte Carlo Methods. *Journal of the Royal Statistical Society, Series B*, 55, 3-23.

U.S. Department of Health and Human Services (1989), The Area Resource File (ARF) System. ODAM Report No. 7-89.

U.S.Department of Health and Human Services, Public Health Service. (1990) *Healthy People 2000: National Health Promotion and Disease Prevention Objectives*. DHHS Publication No. (PHS) 91-50213. Washington: U.S. Government Printing Office.

Wong, G.Y. and Mason, W.W. (1985) The Hierarchical Logistic Regression Model for Multilevel Analysis. *Journal of the American Statistical Association*, 80: 513-524.

Woodruff, R.A.(1966) Use of a Regression Technique to Produce Area Breakdowns of the Monthly National Estimates of Retail Trade. *Journal of the American Statistical Association*, 61: 497-504.

CHAPTER 9

Estimation of Median Income for 4-Person Families by State

Robert Fay and Charles Nelson, U.S. Bureau of the Census
Leon Litow, Department of Health and Human Services

9.1 Introduction and Program History

Starting with income year 1974, the U.S. Census Bureau has computed model-based estimates of median annual income for 4-person families by state using data from the decennial censuses, the Current Population Survey (CPS), and estimates of per capita income (PCI) from the Bureau of Economic Analysis (BEA). Originally, these estimates were used in determining eligibility for the former Title XX Program of the Social Security Act, which provided social services for individuals and families.

Beginning in fiscal year (FY) 1982, the Department of Health and Human Services (HHS) has employed the estimated 4-person family medians to administer the Low Income Home Energy Assistance Program (LIHEAP). This program is one of six block grant programs authorized by the Omnibus Budget Reconciliation Act of 1981 (PL 97-35) and administered by HHS. The Augustus F. Hawkins Human Services Reauthorization Act of 1990 (PL 101-501) reauthorized the LIHEAP through FY 1994.

States, the District of Columbia, Indian tribes and tribal organizations, and territories that wish to assist low income households in meeting the costs of home energy may apply for a LIHEAP block grant. "Home energy" is defined by the LIHEAP statute as "a source of heating or cooling in residential dwellings."

Section 2603(7) of Title XXVI of PL 97-35 requires the Secretary of HHS to establish the state median incomes for purposes of the program. Section 2605(b)(2)(B)(ii) of PL 97-35 provides that 60% of the state median income is one of the income criteria that states can use in determining a household's eligibility for the LIHEAP.

HHS publishes the estimated 4-person family medians by state annually in the *Federal Register*. For purposes of administration, state median incomes are established for families of other sizes as a fixed proportion, depending on the size of family, of the estimated median for 4-person families. The following percentages of the 4-person family medians are used: 52% for 1-person households, 68% for 2 persons, 84% for 3 persons, 100% for 4 persons, 116% for 5 persons, and 132% for 6 persons. For families with more than 6 persons, each person beyond 6 adds an additional 3%. U.S. Bureau of the Census (1991) provides further details, as does the *Federal Register* on March 3, 1988 at 53 FR 6824.

In addition to their programmatic use in the administration of the LIHEAP and the earlier Title XX Program of the Social Security Act, the estimates represent the only intercensal state-specific family income estimates produced by the Census Bureau. Consequently, these estimates have been of interest to a number of general data users. Until the recent publication of the historical series in U.S. Bureau of the Census (1991), however, the estimates did not appear in a regular publication series of the Census Bureau.

9.2 Program Description, Policies, and Practices

Throughout this period the methodology has relied on three sources:

1. Estimates of median family income by state from the decennial censuses. Since the census asks income during the previous year, the census medians pertain to income years 1969, 1979, etc. Although the estimates are based on the long-form sample, the size of this sample provides estimates with virtually negligible sampling errors at the state level every 10 years.

2. Sample estimates of median income by family size by state from the March CPS. Although the CPS estimates are available annually, their direct use is limited by substantial sampling variability due to the size of the CPS sample.

3. Annual estimates of PCI from BEA. These estimates, based on aggregate statistics on components of income from administrative series, have negligible sampling error. The PCI estimates are measures of average income per person, however, and so are only indirectly linked to median income for families.

U.S. Bureau of the Census (1991) describes each of these series in detail. In brief, however:

1.   The decennial census provides geographically detailed estimates by sampling roughly 1/6 of the households in the U.S. to receive the long-form census questionnaire. The census income concept includes as sources: wages and salaries, self-employment income (including losses), Social Security, Supplemental Security Income (SSI), cash public assistance, interest, dividends, rents, royalties, estates and trusts, veterans' payments, unemployment and workers' compensations, private and government survivor, disability, and retirement pensions, alimony, child support, and any other source of money income that is regularly received. Capital gains (or losses) and lump-sums or one-time payments such as life insurance settlements are excluded. Noncash benefits, such as government noncash transfers (foods stamps, Medicaid, etc.) and private sector in-kind benefits are also excluded from the money income definition.

2.   The CPS is a monthly labor force survey of about 60,000 households across the country. Each March, the CPS asks additional questions about money income during the previous year, using the same concepts as the decennial census. The primary purpose of the CPS sample is for national estimates. For example, the CPS provides a national estimate of median income for 4-person families, which is published annually along with many other estimates from the survey.

3.   The BEA income series is based on a different concept of income than the one used by the Census Bureau in the decennial censuses and the CPS. The major difference is that BEA personal income attempts to represent income from all sources, noncash as well as cash. (Appendix A of U.S. Bureau of the Census, 1991, compares the BEA personal income concept to the census money income concept underlying both the CPS and the decennial censuses. Budd, Radner, and Hinrichs, 1973, provide additional detail on this point.)

Another conceptual difference is that the BEA estimates of PCI represent the ratio of estimates of aggregate income to the number of persons in each state. They are not disaggregated by size of family and do not distinguish the income of family members from unrelated individuals and persons in 1-person households.

The PCI series is developed from a variety of government statistics, including Federal tax records from the Department of the Treasury, the insurance files of the Social Security Administration, and state unemployment records collected by the U.S. Department of Labor. The BEA produces annual estimates of personal per capita income for states and other geographic areas. Thus, the BEA estimates generally do not have associated sampling errors, unlike the CPS estimates, since they essentially do not employ sampling techniques. These estimates are described by Bailey, Hazen, and Zabronsky in Chapter 3 of this report.

Before outlining the elements of the methodology, we first compare the estimates for income year 1989, based on the March 1990 CPS and published in March, 1991, with medians for 4-person families from the 1990 census. Figure 9.1 shows the geographic distribution of the true increase in median income for 4-person families during the decade, since the 1980 census.

Figure 9.1 indicates that the greatest relative increase during the decade in the median income of 4-person families occurred in the Northeast region, where most states more than doubled their medians, according to the census. Other areas of active increase include additional states in the East and South Atlantic, and Tennessee, Minnesota, California, and Hawaii. Figure 9.1 also shows that median income in some areas of the country has grown considerably more slowly.

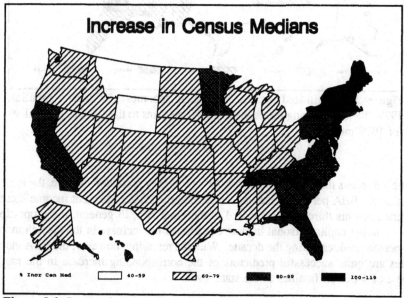

Figure 9.1 Percent increase in census median income for 4-person families, 1979-1989.

Figure 9.2 presents the estimated increase since the 1980 census according to the model. Although there are some differences between the census and the model predictions, the comparison of the two maps shows that the model is successful in capturing most real sources of change in median family income. Some of the states are not classified into the same grouping in Figures 9.1 and 9.2, but, in each case, the difference is by at most one category. For example, states estimated to be among the fastest growing group were either in that category or the next one down, and so forth.

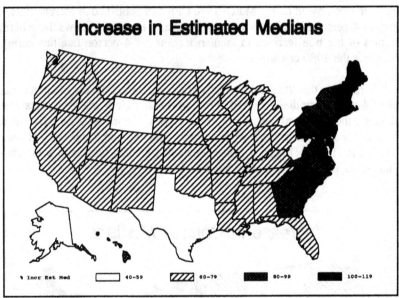

Figure 9.2 Estimated increase in median incomes of 4-person families, 1979-1989, comparing the estimated 1989 medians to the 1980 census values for 1979 medians.

Figure 9.3 shows the geographic distribution of the key predictor variable, the increase in estimated BEA per capita income. Note that the scale of percent income increase is shifted on this third map compared to the other two; in general, the proportional increase in per capita personal income outstripped the increase in the median income of 4-person families during the decade. With the rescaling, however, the BEA income figures are quite successful predictors of the corresponding increase in the median income of 4-person families at the state level.

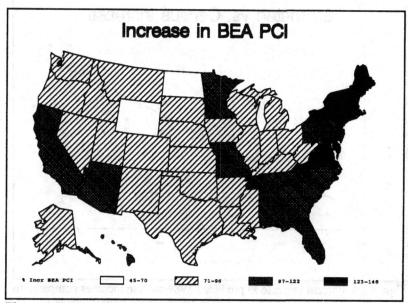

Figure 9.3 Percent increase in BEA PCI for 1979-1989.

Figures 9.4 and 9.5 illustrate additional features of the performance of the model. Both figures include a regression line from the simple linear regression as an aid in assessing fit, although each line is not formally part of the model.

Figure 9.4 compares the estimated increase with the actual increase in median incomes, according to the census. Again, the predictions are not perfect but, nonetheless, appear to capture most of the variation among states in the increase in median income. Figure 9.5 shows that the relationship between increase in BEA PCI and increase in the census median income is essentially linear over the entire spectrum. As previously indicated by the scaling Figure 9.3, Figure 9.5 provides further evidence of somewhat greater dispersion in the increase in the BEA estimates than in the census medians.

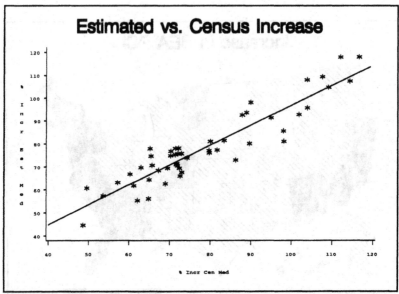

Figure 9.4 Percent increase in predicted 1989 median incomes compared to the 1990 census values, both relative to 1980 census 1979 medians.

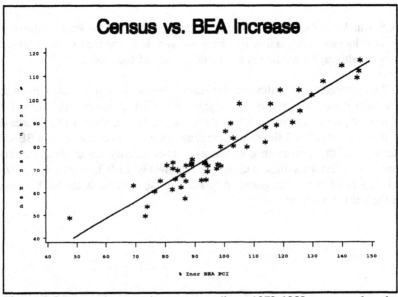

Figure 9.5 Percent increase in census medians, 1979-1989, compared to the percent increase in BEA PCI.

The figures provide a summary of the basic features and performance of the model, and they may be of help to many readers. The remainder of this chapter aims specifically toward a technical audience interested in the exact form of the model, plans for further assessment, and a brief description of potential enhancements. U.S. Bureau of the Census (1991) furnishes a more detailed history of the program, the estimates for calendar years 1974-1989, and citations for the publication of the estimates in the *Federal Register* annually since 1983.

The current methodology has been in place since income year 1984, although with minor refinements over this period of time. The methodology is applied separately for each year, t, in the series. (For simplicity, the implicit subscript, t, is not shown in the following, except where necessary to avoid confusion. Section 9.5 will discuss the possibility of alternative models more attuned to the longitudinal nature of the problem.) The primary elements of the current methodology are:

1) For each state, s, (and the District of Columbia), a direct sample estimate, $\hat{Y}_{s4}$, of the median income for 4-person families is estimated from the CPS. The medians actually are obtained by linear interpolation using tabulated income categorized into intervals of \$2,500. (The census medians were also estimated by interpolation of categorized data.)

2) Similarly, median incomes for 3- and 5-person families, $\hat{Y}_{s3}$ and $\hat{Y}_{s5}$, are estimated as well. For each state, the weighted combination of the two medians,

$$\hat{Y}_{sc} = .75\,\hat{Y}_{s3} + .25\,\hat{Y}_{s5}$$

is computed. The weights, .75 and .25, are approximately proportional to the respective sample sizes, in other words, there are roughly 3 times as many 3-person families as 5-person families.

3) Regressions are fitted to $\hat{Y}_{s4}$ and $\hat{Y}_{sc}$, with separate predictors and corresponding coefficients for each of these two variables. The regressions produce fitted values, $\hat{Y}_{(REG),s4}$ and $\hat{Y}_{(REG),sc}$. The regression model for medians of 4-person families employs 3 predictor variables:

   a) $X_{s41} = 1$, to correspond to a constant term in the model.

b) $X_{s42} = (BEA_{st}/BEA_{sb})\,Y_{(CEN),s4}$, w h e r e BEA_{st} represents BEA PCI for the same income year, t, as $\hat{Y}_{s4}$, and BEA_{sb} and $Y_{(CEN),s4}$ represent BEA PCI and census median income for 4-person families, respectively, for the same base income year, b, of the previouscensus. This predictor variable thus represents the census median adjusted by the proportional growth in BEA PCI since the previous census.

c) $X_{s43} = Y_{(CEN),s4}$, that is, median incomes from the previous census.

The regression model for the weighted average, $\hat{Y}_{sc}$, uses an a n a l o g o u s   s e t   o f   v a r i a b l e s, $X_{sc1} = 1$, $X_{sc2} = (BEA_{st}/BEA_{sb})\,Y_{(CEN),sc}$, and $X_{sc3} = Y_{(CEN),sc}$.

4)   A composite estimate, $\hat{Y}_{(COMP),s4}$, is formed from $\hat{Y}_{s4}$, $\hat{Y}_{sc}$, $\hat{Y}_{(REG),s4}$, and $\hat{Y}_{(REG),sc}$. The combination of the direct sample estimate for 4-person families, $\hat{Y}_{s4}$, with the regression estimate for 4-person families, $\hat{Y}_{(REG),s4}$, is a feature that has appeared in other small domain estimation models based on empirical Bayes procedures. The methodology is in fact multivariate, in using further information present in $\hat{Y}_{sc}$ and $\hat{Y}_{(REG),sc}$ to estimate medians for 4-person families.

Figure 9.5 motivates the inclusion of both $X_{s42} = (BEA_{st}/BEA_{sb})\,Y_{(CEN),s4}$, and $X_{s43} = Y_{(CEN),s4}$, as predictors in the model. Given the consistent linear relationship between proportional change in BEA PCI and change in the census median, the first of these two terms is the most obvious single expression of this relationship. The second of the two complements the first: without the second, the regression would be satisfactory only if the slope in Figure 9.5 were 1.0. In fact, the slope is somewhat less than 1. Inclusion of both predictors allows, in effect, the slope of the regression line in Figure 9.5 to be estimated from the CPS data.

A key feature of the model is the multivariate combination of estimation of the target variable of interest, median income for 4-person families, along with an auxiliary variable, the combined 3- and 5-person family medians, even though the auxiliary variable is not itself a subject of interest. In fact, the purpose of the multivariate approach is to realize additional gains in the estimation of 4-person family medians. Fuller and Harter (1987) and Fay (1987) motivate the possible advantages of the multivariate approach for problems of this sort.

The current model replaced an earlier version, whose major features were:

1) The earlier model began with the CPS medians for 4-person families, $\hat{Y}_{s4}$, but did not consider medians for other sizes of families.

2) Predictors $X_{s41} = 1$ and $X_{s42} = (BEA_{st}/BEA_{sb})\,Y_{(CEN),s4}$ were employed as now, but $X_{s43} = Y_{(CEN),s4}$ was not included in the regression.

3) A regression was fitted to $\hat{Y}_{s4}$, with the two predictors from step 2).

4) For each state, s, a weighted composite, $\hat{Y}_{(COMP),s4}$, of $\hat{Y}_{(REG),s4}$ and $\hat{Y}_{s4}$ was formed.

5) The final estimate, $\hat{Y}_{(FIN),s4}$, was derived from $\hat{Y}_{(COMP),s4}$ by constraining it to lie within one standard deviation of $\hat{Y}_{s4}$. In other words, if $\sigma_{(\hat{Y}),s4}$ represents the standard error of $\hat{Y}_{s4}$, then

$$\hat{Y}_{(FIN),s4} = \hat{Y}_{s4} + \sigma_{(\hat{Y}),s4} \quad \text{if } \hat{Y}_{(COMP),s4} > \hat{Y}_{s4} + \sigma_{(\hat{Y}),s4}$$

$$= \hat{Y}_{(COMP),s4} \quad \text{if } \hat{Y}_{s4} - \sigma_{(\hat{Y}),s4} < \hat{Y}_{(COMP),s4} < \hat{Y}_{s4} + \sigma_{(\hat{Y}),s4}$$

$$= \hat{Y}_{s4} - \sigma_{(\hat{Y}),s4} \quad \text{otherwise}$$

Section 9.4.1 discusses features and properties of this earlier estimator.

9.3 Estimator Documentation

9.3.1 General Features of the Current Estimator

The last two steps, 3) and 4), of the current procedure can be represented more completely in matrix notation. Let $\hat{Y}$ represent a 102-element column vector formed from the CPS sample medians in each state, i.e.,

$$\hat{Y} = (\hat{Y}_{1,4}, \hat{Y}_{1,c}, \hat{Y}_{2,4}, \hat{Y}_{2,c}, \cdots, \hat{Y}_{51,4}, \hat{Y}_{51,c})'$$

Let $X_{s4} = (X_{s41}, X_{s42}, X_{s43})$ denote a row vector with the 3 predictor variables for the 4-person family median in state s, and $X_{sc} = (X_{sc1}, X_{sc2}, X_{sc3})$ the predictors for the weighted combined variable, $\hat{Y}_{sc}$, and let

$$X = \begin{pmatrix} X_{1,4} & 0 \\ 0 & X_{1,c} \\ X_{2,4} & 0 \\ 0 & X_{2,c} \\ \cdots & \cdots \end{pmatrix}$$

be a 102 by 6 matrix containing all the predictors. The small domain estimator is based on the model:

$$\hat{Y} = X\beta + b + e$$

where β represents a 6 by 1 column vector of the regression coefficients, b represents a 102 by 1 column vector of random effects denoting the departure of the individual true medians from the regression predictions, and e denotes a 102 by 1 column vector of sampling errors. The covariance matrix of b is assumed to be in a block diagonal form:

$$A^* = \begin{bmatrix} A & 0 & 0 & \cdots \\ 0 & A & 0 & \cdots \\ 0 & 0 & A & \cdots \\ \cdot & \cdot & \cdot & \cdots \end{bmatrix}$$

where A represents a 2 by 2 covariance matrix for the model errors. Note that the model assumes that the distribution of the random effects, b, is identical in each state and uncorrelated between states. The sampling covariances of e are assumed to be of the form

$$D^* = \begin{bmatrix} D_1 & 0 & 0 & ... \\ 0 & D_2 & 0 & ... \\ 0 & 0 & D_3 & ... \\ . & . & . & ... \end{bmatrix}$$

where D_s is a 2 by 2 covariance matrix for the sampling errors for state s. Consequently, unlike the assumption that the random effects have the same distribution in each state, D_s is allowed to vary over states. Further comments on the derivation of A and D_s follow in sections 9.3.3 and 9.3.2, respectively.

Given A^* and D^*, the best linear unbiased estimate (BLUE) of β is

$$\hat{\beta} = (X'(D^*+A^*)^{-1}X)^{-1}X'(D^*+A^*)^{-1}\hat{Y},$$

and the BLUE of the true medians,

$$Y = X\beta + b$$

is

$$\hat{Y}_{(COMP)} = X\hat{\beta} + A^*(D^*+A^*)^{-1}(\hat{Y} - X\hat{\beta}) \tag{9.1}$$

Because of the block diagonal forms of A^* and D^*, this estimator forms a weighted average of the sample and regression estimates within each state for both the 4-person and combined family medians, as described earlier at 4).

Estimator (9.1) has been employed for income years 1984-1991, but refinements have entered in the estimation of the sampling errors, D_s, and model errors, A.

9.3.2 Estimation of the Sampling Errors of the CPS Medians

As noted earlier, the CPS sample estimates, $\hat{Y}_{s3}$, $\hat{Y}_{s4}$, and $\hat{Y}_{s5}$, were obtained from linear interpolation of the estimated income distribution in intervals of $2,500. Woodruff (1952) established a general approach for estimating the variance of an estimated median by computing the variance of the proportion less than the estimated median, when the estimated median is treated as fixed, and dividing this variance by the square of an estimate of the density (that is, the first derivative of the distribution function) at the median.

Although Woodruff's original method was based on computing the density of the distribution in the interval containing the estimated median, the small expected number of sample cases falling into each of the $2,500 intervals in the estimated CPS income distribution by family size within states leads this approach to be unstable. Empirically, experimentation with samples of 100 and 400 cases drawn from the national income distribution showed that it was preferable to use a broader interval to estimate the density for samples of these sizes. Specifically, in addition to the interval containing the median, the 4 $2,500 intervals immediately below and 2 intervals immediately above are used to form a combined interval of width $17,500 for purposes of estimating the density, d. The variance estimator is:

$$\hat{V}(\hat{Y}_{s4}) = \frac{\hat{V}(p_m)}{d^2} \tag{9.2}$$

where $\hat{V}(p_m)$ represents the variance of the estimated proportion less than $\hat{Y}_{s4}$, and d = (the proportion in the interval)/$17,500.

Originally, $\hat{V}(p_m)$ was estimated by extrapolation from results of CPS variance estimates for national characteristics in the early 1970's. These results in large part formed the basis of the variance generalizations frequently provided in Census Bureau reports (for example, U.S. Bureau of the Census 1993, Appendix D). For internal use, Census Bureau staff interpolated the parameters of the variance generalization from the national to the state level under assumptions about the effect of sample size and about the contributions of variance due to the first-stage selection of primary sampling units in the CPS design.

In 1989, direct estimation of $\hat{V}(p_m)$ became possible for the 1988 estimates. Comparison of the direct estimates with the generalized variances indicated that the latter appeared to underestimate the state variances by approximately 20%, although there was not a discrepancy this large at the national level. The discrepancy indicated probable shortcomings of the extension of the variance generalization to the state level. The difficulty in generalizing to the state level possibly arose because the benefits of the complex ratio estimation employed in the CPS, which realizes variance reductions for income statistics at the national level, probably does not extend to the state estimates. Thus, the generalization was too optimistic at the state level, in hindsight.

The replicate weights used to compute variances for the 1988 estimates were not consistently available for CPS samples in other years, at the time. Consequently, it was of interest to find a second, even if more approximate, method to compute variance estimates directly. Thus, a second comparison was of interest: the relationship between the direct estimates using replicate weights and simpler variance estimates computed from variation among sample segments within sampled primary sampled units. The latter calculation estimates "within" variance, the component of variance due to sampling segments of housing units within sampled primary sampling units. In other words, the within variance is a conditional variance, given the selection of primary sampling units. The estimated variance based on the replicate weights, which included all components of variance, was in close agreement, on the average, with the within variance estimated from variation between segments. In addition, the estimated within variance is a theoretically more stable estimate.

Consequently, direct calculation of $\hat{V}(p_m)$ based on within variance has replaced variance generalization since income year 1988. For the 1988 estimates, a partial variance generalization averaging parameters of the variance model for 3- and 4-person families within each state was employed to smooth the variance estimates somewhat. In subsequent years, $\hat{V}(p_m)$ has been estimated directly for each family size, again at the multiple of \$2,500 closest to the sample median.

Until the availability of direct variance estimates, the non-diagonal elements of D_s were set to zero on the presumption of that the correlation between estimates for different family sizes would be close to zero. Since CPS segments may contain a mixture of households with different size families who are somewhat similar with respect to income, a small positive correlation is possible, however. In 1989, the correlation between 3- and 4-person family medians, on a national level, was estimated at approximately .11, and this correlation was used within each state to estimate D_s. Similar estimates of national correlation have been used in subsequent years.

9.3.3 Estimation of the Model Errors

The remaining element of (9.1) is the estimation of A^*. As noted earlier, A^* is assumed to have a block diagonal form with identical 2 by 2 covariances, A, on the diagonal. In other words, the model errors in each state were assumed to have an identical distribution.

Empirical Bayes estimates characteristically estimate components of model error, such as A, directly from the data. A typical experience, however, is that such estimates are themselves subject to considerable variation, which in turn increases the variance of estimators such as (9.1) compared to the variance of (9.1), if A were known.

By fitting the regression model based on the 1970 census medians and changes in BEA PCI between 1969 and 1979 to the 1980 census values, an estimate of model variance may be obtained from the differences between the resulting predicted and 1980 census values. The results were:

$$A_{(CEN)} = \begin{pmatrix} 0.308 & 0.263 \\ 0.263 & 0.286 \end{pmatrix} \times 10^6$$

Until income year 1989, A has been estimated by projecting $A_{(CEN)}$ for the proportional change in national median incomes by family size. Specifically, if $\hat{Y}_{(NAT),4t}$ represents the national CPS median for year t, $Y_{(CEN),4b}$ the national census median for the census year, b, $r_{4t} = \hat{Y}_{(NAT),4t}/Y_{(CEN),4b}$, and r_{ct} is defined similarly for the combined 3- and 5-person family medians, then

$$A_{(PROJ),t} = \begin{pmatrix} r_{4t}^2 A_{(CEN),11} & r_{4t} r_{ct} A_{(CEN),12} \\ r_{4t} r_{ct} A_{(CEN),21} & r_{ct}^2 A_{(CEN),22} \end{pmatrix}$$

Each year, $A_{(PROJ),t}$ has been compared to the maximum-likelihood estimate, $\hat{A}_{(MLE),t}$. Fay (1986) further describes this comparison and presents values for income years 1981-1984. Subsequently, the maximum-likelihood estimate, $\hat{A}_{(MLE),t}$, tended to stay within sampling error of $A_{(PROJ),t}$ but nonetheless exhibited diagonal elements usually larger than $A_{(PROJ),t}$.

For income years 1989 and 1990, $A_{(PROJ),t}$ has been replaced by

$$\hat{A}_{(MULT),t} = \hat{\lambda}_t A_{(CEN)}$$

where $\hat{\lambda}_t$ is obtained by maximum-likelihood estimation. In other words, this estimator employs a single factor to inflate the increase in model uncertainty since the previous census, in place of estimating three parameters. The results for income year 1989 were:

$$\hat{A}_{(MULT),t} = \begin{pmatrix} 3.154 & 2.698 \\ 2.698 & 2.932 \end{pmatrix} \times 10^6$$

compared to:

$$A_{(PROJ),t} = \begin{pmatrix} 1.000 & 0.827 \\ 0.827 & 0.870 \end{pmatrix} \times 10^6$$

Consequently, the current method suggests considerably greater increase in model error than accounted for by the assumptions underlying the original projection.

In a manner analogous to the calculation for the 1980 census, fitting the regression model directly to the recently available 1990 census results yields the following estimate of model error:

$$A_{(CEN)} = \begin{pmatrix} 1.631 & 1.444 \\ 1.444 & 1.494 \end{pmatrix} \times 10^6$$

This outcome is between the multiplicative results and the projection, although closer to the latter. In fact, much of the increase in the estimate of error based on the CPS sample estimates may be attributed to greater disagreement between the 1990 CPS and census state estimates of median income than expected based on CPS sampling variability, and further remarks on this point will be included in the final section. Note also how strikingly well these empirical results fit the multiplicative model. For example, choosing $\lambda_t = 5.2955$ gives:

$$A_{(MULT)} = \begin{pmatrix} 1.631 & 1.394 \\ 1.394 & 1.514 \end{pmatrix} \times 10^6$$

9.4 Evaluation Practices

9.4.1 Comparisons to the 1980 Census

The availability of direct census estimates every 10 years affords a significant opportunity to evaluate and recalibrate the estimation technique. The current methodology grew from its predecessor, described at the end of section 9.2, primarily as a consequence of comparisons to 1980 census results. The conclusions of that comparison, discussed in Fay (1986), were, in brief:

1) The earlier method yielded generally useful state estimates, but 1 estimate was in error by more than 10% and 11 additional estimates were in error by 5% or more.

2)   The constraint that the final estimate, $\hat{Y}_{(FIN),s}$, lie within one standard error of the sample estimate actually yielded an estimator that was somewhat less effective overall without actually realizing a distinct benefit. This procedure, suggested by Efron and Morris (1971) and used, for example, by Fay and Herriot (1979) and others, is advantageous when a few domains have true values that deviate by relatively large amounts from the model predictions. The essentially normal distribution of model errors in 1980 favored dropping the constraint, however.

3)   The predictor $X_{s43} = Y_{(CEN),s4}$ was added to the model because the estimates of relative change based on the BEA PCI estimates appeared overly disperse relative to the actual census changes. As noted earlier, the inclusion of $X_{s43} = Y_{(CEN),s4}$ allows the slope in Figure 9.5 to differ from 1.0.

The current model was developed by testing against the 1980 census data, using the 1970 census as the base. In particular, this evaluation provided $A_{(CEN)}$ used in the model through the 1990 estimates. In estimating the 1984-1990 medians, the 1980 census figures were used as the base. The 1990 census figures were first used as the base in estimating 1991 medians.

9.4.2 Comparisons to the 1990 Census

Figures 9.1 - 9.5 compare the model to recently available estimates from the 1990 census. Overall, the results of the comparison are quite encouraging. For example, no estimate was in error by 10% or more, and only 7 were in error by 5% or more. These findings reflect only the first steps in a more complete analysis.

The next critical step, however, will be to react to a surprising finding reported in Section 9.3.3, namely, that the CPS sample estimates of the medians by state appear to differ from the CPS values by more than sampling error alone would suggest. This is in contrast to the comparison of the 1980 CPS and census. Consequently, some form of nonsampling error is possible, but a more systematic study of components of differences between the CPS and the census will be required to isolate the significant source or sources of these differences.

Figures 9.1 and 9.2 provide one suggestion of a possible source of difference: relative to the census, the model underpredicted increase in four large states: California, Florida, New York, and Texas. In each of these states, the CPS sample estimates themselves fall below the census values. All four states also have appreciable

Hispanic populations. Furthermore, preliminary comparisons suggest higher estimated medians for Hispanics in the census compared to the CPS.

Hispanic income may be only one of several factors underlying differences between census and CPS state medians. The outcome of a more complete investigation should provide a firmer basis to separate the issues of limitations of the model from possible nonsampling error in either the CPS or the census.

Once issues of nonsampling error are more firmly understood, the census results should permit assessment of a number of features of the current model:

1) The average error of the model predictions.

2) Whether errors are differential for certain classes of states, e.g., small vs. large, rapidly changing vs. static, lower income vs. higher, etc.

3) Whether errors cluster geographically.

4) Whether modification of the current predictors would yield significant improvement in prediction.

The census data permit assessment of the current model but also offer the occasion for consideration of more significant changes for subsequent years. A number of these are described in the next section.

In addition to relying on the census for evaluation, work on alternative models, such as the hierarchical Bayes model described by Datta, Fay, and Ghosh (1991), has addressed methods to obtain estimates of individual and aggregate measures of performance from the sample estimates when census data are not available. The 1990 census data should help to calibrate these procedures for future use. (Unfortunately, these procedures may be adversely affected by nonsampling error producing differences between the expected values of the CPS and the census medians at the state level, so that understanding sources of nonsampling error is a critical step here as well.)

9.5 Current Problems and Activities

Implemented a year at a time, the current model and its predecessor has produced a series spanning income years 1974 to 1991 without taking any advantage of the longitudinal or time series nature of this problem. Several researchers have investigated longitudinal extensions that attempt to address this aspect.

The current model relies simply on observed relationships that appear to be quite linear, without taking advantage of any specific knowledge about income distributions. Possibly, a more explicit parametric model for the income distribution may represent a fruitful alternative. On the other hand, the utility of such efforts would have to be balanced against requirements of parsimony imposed by the relatively small sizes of the CPS state samples.

As noted at the end of Section 9.4, another area of potential research is to attempt to improve measures of error from the model for the intercensal period. Recent research in fully Bayes procedures may prove promising for estimation of error.

REFERENCES

Budd, E. C., Radner, D. B., and Hinrichs, J. C. (1973), "Size Distribution of Family Personal Income: Methodology and Estimates for 1964," *BEA-SP 73-21*, Washington, DC: Bureau of Economic Analysis.

Datta, G., Fay, R. E., and Ghosh, M. (1991), "Hierarchical and Empirical Multivariate Bayes Analysis in Small Area Estimation," in *Proceedings of the Annual Research Conference*, Washington, DC: U. S. Bureau of the Census, pp. 63-79.

Efron, B., and Morris, C. (1971), "Limiting the Risk of Bayes and Empirical Bayes Estimators -Part I: the Bayes Case," *Journal of the American Statistical Association*, 74, 269-277.

Fay, R. E. (1986), "Multivariate Components of Variance Models as Empirical Bayes Procedures for Small Domain Estimation," in *Proceedings of the Survey Research Methods Section*, Washington, DC, American Statistical Association, pp. 99-107.

_____ (1987), "Application of Multivariate Regression to Small Domain Estimation," in *Small Area Statistics, An International Symposium*, R. Platek, J. N. K. Rao, C. E. Särndal, and M. P. Singh, eds., New York: John Wiley & Sons, pp. 91-102.

Fay, R. E. and Herriot, R. A. (1979), "Estimates of Income for Small Places: An Application of James-Stein Procedures to Census Data," *Journal of the American Statistical Association*, 74, 269-277.

Fuller, W. A. and Harter, R. M. (1987), "The Multivariate Components of Variance Model for Small Domain Estimation," in *Small Area Statistics, An International Symposium*, R. Platek, J. N. K. Rao, C. E. Särndal, and M. P. Singh, eds., New York: John Wiley & Sons, pp. 103-123.

U.S. Bureau of the Census (1991), "Estimates of Median 4-Person Family Income by State: 1974-1989," Current Population Reports, Technical Paper No. 61, U.S. Government Printing Office, Washington, DC.

_____ (1993), "Money Income of Households, Families, and Persons in the United States," Current Population Reports, Consumer Income Series, P60, No.194, U.S. Government Printing Office, Washington, DC.

Woodruff, R. (1952), "Confidence Intervals for Medians and Other Position Measures," *Journal of the American Statistical Association*, 47, 635-646.

CHAPTER 10

Recommendations and Cautions

During the design of a data system, indirect estimators rarely, if ever, are considered for Federal statistical programs when resources to produce direct estimates of adequate precision are available. However, given an existing system, if direct estimation is judged to be inadequate for a domain not specified in the design, indirect estimation may, in some cases, prove to be a valuable alternative. There are a number of reasons that direct estimators are preferable to indirect ones and, if federal statistical agencies are to improve the usefulness of indirect estimates, a number of important issues, including those that follow, should receive additional attention. The following points are developed further in the sections of this chapter.

• Traditionally, statistical programs are designed with only direct estimators for large domains in mind; indirect estimators for small domains are considered only after the design has been determined. Planning for both direct estimators and indirect estimators at the sample design stage should lead to improved indirect estimates.

• The purpose of the analysis should be kept in mind when selecting an indirect estimator. This is important with direct estimators, but even more so with indirect estimators where there may be implications for the choice between a domain indirect and a time indirect estimator.

• More coordination and cooperation among Federal agencies would increase accessibility to auxiliary information for use with indirect estimators.

• Additional empirical evaluations of existing and proposed indirect estimators are needed. Existing evaluations are generally limited in the conclusions they are able to draw.

- Additional research on errors associated with indirect estimators is necessary. Additional attention should be directed to the estimation, not only of variances, but also of biases, mean square errors, and confidence intervals.

- When indirect estimates are published, they should be distinguished from direct estimates and accompanied by a clear explanation of model assumptions and appropriate cautions.

10.1 Sample Designs for Small Areas

The design of samples for the production of both direct and indirect estimates should receive more attention.

Since the 1940's, considerable research has been conducted on the design of samples for use with direct estimators. More recently, indirect estimators have received the attention of researchers, but most often in a way that treats the sample design as fixed and beyond the scope of the research effort. An exception to this statement is found in the work on rolling samples and censuses (for example, Kish 1990 with discussion by Scheuren 1990 and Hansen 1990) where direct and/or indirect estimators may be defined depending on the population quantity being estimated and whether the design requires a subset of units to be observed in more than one time period of interest.

Present applications of indirect estimators generally evolved from data systems designed for other purposes. Few, if any, existing data sources were created with the production of indirect estimates as a consideration. Sample designs considering the production of both direct estimates and indirect estimates deserve much more attention than they have received thus far. This is particularly true for continuing surveys where information useful for design and estimator evaluation is obtained periodically.

None of the programs described in this report were designed with indirect estimation as an explicit consideration. However, indirect estimators which benefit from observations in the domain and time of interest have been enhanced by certain design decisions. For example, the redesign of the National Health Interview Survey, discussed in Chapter 8, will include stratification by individual states so that the sample size within each state is controlled.

10.2 Use of Estimates and the Selection of an Estimator

Selection of an appropriate indirect estimation method should take into account the purpose for which estimates are to be used.

Indirect estimators should be selected with great caution and perhaps even avoided in some situations. Indirect estimators may perform poorly when the purpose of the analysis is to identify domains with extreme population values, to rank domains, or to identify domains that fall above or below some predetermined level.

A domain indirect estimator borrows strength across domains and is justified under a model that assumes model parameters are the same across domains. If the purpose of the analysis is to make comparisons across domains for a given time period, an inconsistency between objective and method can be avoided if a time indirect estimator is chosen. Of course, depending on the application and the available auxiliary information, this may not always be the appropriate course of action. Similarly, if the purpose of the analysis is to make comparisons across time periods within a given domain, it may be more appropriate to select from among domain indirect estimators.

10.3 Auxiliary Information

More coordination and cooperation among Federal agencies would allow expanded access to the auxiliary information on which indirect estimators depend.

Regardless of how appropriate the conceptual and theoretical basis of a particular indirect estimator may be, the estimator cannot be used in practice if the required auxiliary variables, which usually come from administrative sources or censuses, are not available. Without auxiliary information related to the variable of interest and for the domain and time period of interest, only the most crude indirect estimators can be implemented.

For programs described in this report, the search for auxiliary variables generally seems to have been somewhat ad hoc, with little coordination or cooperation among statistical agencies. An integrated data system for geographical areas would make auxiliary information more readily available and would potentially lead to improved indirect estimators. Such a system might also take advantage of recent computational technologies. Although not previously discussed in this report and designed with the objectives of mental health needs assessment, policy, and research in mind, the National Institute of Mental Health's Health Demographic Profile System (Goldsmith et al 1984) provides a variety of social indicators for geographic areas. In addition, Statistics Canada has addressed this issue in their Small Area Data Program (Brackstone 1987).

10.4 Empirical Evaluations

Additional empirical evaluations are needed to help determine whether indirect estimators are adequate for the intended purposes.

The decision whether or not to use an indirect estimator is rarely an easy one. Empirical evaluations play a critical role in the decision process. The performance of an indirect estimator in a given application depends on the variable(s) of interest and their relationship to the auxiliary variables through the underlying model. Generalization from one application to another is difficult so that each application requires a different empirical evaluation.

In practice, indirect estimators are considered for use in situations where data are not available to support the use of direct estimators. The data that, if available, would support the use of direct estimators are the same data that would be most useful in the evaluation of indirect estimators and models. In other words, the need for indirect estimators is the greatest in precisely those situations where data are not available for their adequate empirical evaluation. For this reason, it is rare that a single empirical evaluation of an indirect estimator is completely convincing. There seems to be no satisfying solution to this problem. Resourcefulness in locating data sources and the use of multiple empirical evaluations will be required in most, if not all, situations.

Two approaches can be used for empirical evaluations of indirect estimators. In the first, estimators under consideration are used to produce estimates; these estimates then are compared to a better estimate or census value. The estimator that performs best using an empirical average squared error or similar criterion is judged to be most appropriate for the given application. The great majority of evaluation efforts connected with indirect estimates have used this approach.

The second approach is to evaluate how well the models associated with the competing estimators fit the data. A principled approach is needed; models should not only fit the data but also have a conceptual basis. An indiscriminate search through a large number of models does not often produce appealing results.

Empirical evaluations of indirect estimators are critical, and careful evaluations should include consideration of underlying models as well as the corresponding estimators. In addition to initial evaluations leading to the selection of an estimator, continuing evaluations of the underlying models should be conducted for those series that are published periodically.

10.5 Measures of Errors for Indirect Estimators

Care should be taken in the production of measures of errors of indirect estimators. Estimates of variances alone may be misleading to data users. Additional research on the estimation of biases, mean squared errors, and confidence intervals is needed.

At present, none of the programs described in this report provide measures of error to accompany published indirect estimates. It is difficult to produce meaningful measures of error for indirect estimators. Expressions for indirect estimator variances and biases under the assumed model are usually straightforward to derive, and estimation of variances is usually possible. If the model leading to an estimator is a good approximation of reality in a given application, then the variance of the estimator derived under the model should serve as an adequate measure of error. However, if the model associated with the estimator is not a good approximation, the estimator will have a bias due to model failure. If the bias is large relative to the variance, the variance, by itself, will not be an adequate measure of error, and an estimate of the mean squared error will be required. This is a difficult problem since an estimate of the mean squared error requires an estimate of the bias. Bias in an indirect estimator arises from model failure; that is, failure of the model to adequately represents the variability of the variable of interest over domains and time. Since the population quantity being estimated is specific to a given domain and time, it follows that an estimate of this bias requires data from that domain and time. If the available data are inadequate to produce reliable direct estimates, it is unlikely that they would be adequate to support acceptable estimates of biases. Estimation of confidence intervals for indirect estimators is also a difficult problem in practice. The existing research in this area provides valuable results, but additional work is needed. In the interim, measures of error as indicated by empirical evaluation studies may be the only source of error information for users.

10.6 Publication of Indirect Estimates

A clear distinction should be made between direct and indirect estimators. When indirect estimates are published, they should be accompanied by appropriate cautions and clear explanations of the model assumptions.

Direct estimates published by Federal statistical agencies usually meet expected reliability and validity criteria. Even unsophisticated users of statistics have come to expect estimates from Federal statistical agencies to be trustworthy in some sense. Rarely is enough known about the error structure of indirect estimators to produce adequate measures of their quality. For this reason, it is misleading to the public and potentially damaging to the reputation of Federal statistical agencies to publish indirect estimates that are not clearly distinguished from direct estimates and that are not offered with appropriate cautions. In any case, a clear statement of the assumptions required for the indirect estimator to be model unbiased should be included with all published estimates. This issue has been addressed differently by various programs. For example, the Bureau of Labor Statistics produces estimates for a limited number of states using a direct estimator and for the remainder of the states using an indirect estimator. The two sets of estimates are published in the same table but separated into the two groups with explanatory notes. The National Center for Health Statistics publishes indirect estimates from the National Health Interview Survey in a separate publication containing explanations and cautions.

10.7 Cautions for Producers and Users of Indirect Estimates

As evidenced by the large and growing literature on indirect estimation methods, numerous researchers have been working on the challenging problems facing those who must produce estimates with inadequate resources. Many authors suggest new approaches or variations of existing approaches, but few caution about the dangers associated with the use of indirect estimation methods. The following exceptions should be noted.

"The synthetic estimator is a dangerous tool, but with careful further development, it has an attractive potential." (Simmons 1979)

"A workshop of this sort, focused on a specific technique, can spur development, but it can also be dangerous. The danger is that, from hearing many people speak many words about synthetic estimation we become comfortable with the technique. The idea and the jargon become familiar, and it is easy to accept that 'Since all these people are studying synthetic estimation, it must be okay.' We must remain skeptical and not allow familiarity to dull our healthy skepticism. There is reason for some optimism, but it must be guarded optimism." (Royall 1979)

" . . . a cautious approach should be adopted to the use of small area estimates, and especially to their publication by government statistical agencies. When government statistical agencies do produce model-dependent small area estimates, they need to distinguish them clearly from conventional sample-based estimates. . . . Before small area estimates can be considered fully credible, carefully conducted evaluation studies are needed to check on the adequacy of the model being used. Sometimes model-dependent small area estimators turn out to be of superior quality to sample-based estimators, and this may make them seem attractive. However, the proper criterion for assessing their quality is whether they are sufficiently accurate for the purposes for which they are to be used. In many cases, even though they are better than sample-based estimators, they are subject to too high a level of error to make them acceptable as the basis for policy decisions." (Kalton 1987)

Indirect estimation should be considered when other, more robust alternatives are unavailable, and then only with appropriate caution and in conjunction with substantial research and evaluation. Even after such efforts, neither producers nor users should forget that indirect estimates may not be adequate for the intended purpose.

REFERENCES

Brackstone, G.J. (1987), "Small Area Data: Policy Issues and Technical Challenges," in *Small Area Statistics*, New York: John Wiley and Sons.

Goldsmith, H.G., Jackson, D.J., Doenhoefer, S., Johnson, W., Tweed, D.L., Stiles, D., Barbano, J.P., and Warheit, G. (1984), "The Health Demographic Profile System's Inventory of Small Area Social Indicators," National Institute of Mental Health. Series BN No. 4., DHHS Pub. No. (ADM) 84-1354. Washington, D.C.: U.S. Government Printing Office.

Hansen, M.H., (1990), "Discussion of paper by Kish," *Survey Methodology*, 16-1, 81-86.

Kalton, G. (1987), "Panel Discussion" in *Small Area Statistics*, New York: John Wiley and Sons.

Kish, L., (1990), "Rolling Samples and Censuses," *Survey Methodology*, 16-1, 63-71.

Royall, R.A. (1979), "Prediction Models in Small Area Estimation," in *Synthetic Estimates for Small Areas* (National Institute on Drug Abuse, Research Monograph 24), Washington, D.C.: U.S. Government Printing Office.

Scheuren, F., (1990), "Discussion of paper by Kish," *Survey Methodology*, 16-1, 72-79.

Simmons, W.R. (1979), "Discussion of a paper by Levy," in *Synthetic Estimates for Small Areas* (National Institute on Drug Abuse, Research Monograph 24), Washington, D.C.: U.S. Government Printing Office.

Lecture Notes in Statistics

For information about Volumes 1 to 21
please contact Springer-Verlag

Vol. 65: A. Janssen, D.M. Mason, Non-Standard Rank Tests. vi, 252 pages, 1990.

Vol 66: T. Wright, Exact Confidence Bounds when Sampling from Small Finite Universes. xvi, 431 pages, 1991.

Vol. 67: M.A. Tanner, Tools for Statistical Inference: Observed Data and Data Augmentation Methods. vi, 110 pages, 1991.

Vol. 68: M. Taniguchi, Higher Order Asymptotic Theory for Time Series Analysis. viii, 160 pages, 1991.

Vol. 69: N.J.D. Nagelkerke, Maximum Likelihood Estimation of Functional Relationships. V, 110 pages, 1992.

Vol. 70: K. Iida, Studies on the Optimal Search Plan. viii, 130 pages, 1992.

Vol. 71: E.M.R.A. Engel, A Road to Randomness in Physical Systems. ix, 155 pages, 1992.

Vol. 72: J.K. Lindsey, The Analysis of Stochastic Processes using GLIM. vi, 294 pages, 1992.

Vol. 73: B.C. Arnold, E. Castillo, J.-M. Sarabia, Conditionally Specified Distributions. xiii, 151 pages, 1992.

Vol. 74: P. Barone, A. Frigessi, M. Piccioni, Stochastic Models, Statistical Methods, and Algorithms in Image Analysis. vi, 258 pages, 1992.

Vol. 75: P.K. Goel, N.S. Iyengar (Eds.), Bayesian Analysis in Statistics and Econometrics. xi, 410 pages, 1992.

Vol. 76: L. Bondesson, Generalized Gamma Convolutions and Related Classes of Distributions and Densities. viii, 173 pages, 1992.

Vol. 77: E. Mammen, When Does Bootstrap Work? Asymptotic Results and Simulations. vi, 196 pages, 1992.

Vol. 78: L. Fahrmeir, B. Francis, R. Gilchrist, G. Tutz (Eds.), Advances in GLIM and Statistical Modelling: Proceedings of the GLIM92 Conference and the 7th International Workshop on Statistical Modelling, Munich, 13-17 July 1992. ix, 225 pages, 1992.

Vol. 79: N. Schmitz, Optimal Sequentially Planned Decision Procedures. xii, 209 pages, 1992.

Vol. 80: M. Fligner, J. Verducci (Eds.), Probability Models and Statistical Analyses for Ranking Data. xxii, 306 pages, 1992.

Vol. 81: P. Spirtes, C. Glymour, R. Scheines, Causation, Prediction, and Search. xxiii, 526 pages, 1993.

Vol. 82: A. Korostelev and A. Tsybakov, Minimax Theory of Image Reconstruction. xii, 268 pages, 1993.

Vol. 83: C. Gatsonis, J. Hodges, R. Kass, N. Singpurwalla (Editors), Case Studies in Bayesian Statistics. xii, 437 pages, 1993.

Vol. 84: S. Yamada, Pivotal Measures in Statistical Experiments and Sufficiency. vii, 129 pages, 1994.

Vol. 85: P. Doukhan, Mixing: Properties and Examples. xi, 142 pages, 1994.

Vol. 86: W. Vach, Logistic Regression with Missing Values in the Covariates. xi, 139 pages, 1994.

Vol. 87: J. Møller, Lectures on Random Voronoi Tessellations. v 134 pages, 1994.

Vol. 88: J. E. Kolassa, Series Approximation Methods in Statistics. viii, 150 pages, 1994.

Vol. 89: P. Cheeseman, R.W. Oldford (Editors), Selecting Mod From Data: AI and Statistics IV. xii, 487 pages, 1994.

Vol. 90: A. Csenki, Dependability for Systems with a Partition State Space: Markov and Semi-Markov Theory and Computatio Implementation. x, 241 pages, 1994.

Vol. 91: J.D. Malley, Statistical Applications of Jordan Algebr viii, 101 pages, 1994.

Vol. 92: M. Eerola, Probabilistic Causality in Longitudinal Stud vii, 133 pages, 1994.

Vol. 93: Bernard Van Cutsem (Editor), Classification a Dissimilarity Analysis. xiv, 238 pages, 1994.

Vol. 94: Jane F. Gentleman and G.A. Whitmore (Editors), C Studies in Data Analysis. viii, 262 pages, 1994.

Vol. 95: Shelemyahu Zacks, Stochastic Visibility in Rand Fields. x, 175 pages, 1994.

Vol. 96: Ibrahim Rahimov, Random Sums and Branching Stocha Processes. viii, 195 pages, 1995.

Vol. 97: R. Szekli, Stochastic Ordering and Dependence in App Probablility. viii, 194 pages, 1995.

Vol. 98: Philippe Barbe and Patrice Bertail, The Weigh Bootstrap. viii, 230 pages, 1995.

Vol. 99: C.C. Heyde (Editor), Branching Processes: Proceedi of the First World Congress. viii, 185 pages, 1995.

Vol. 100: Wlodzimierz Bryc, The Normal Distributi Characterizations with Applications. viii, 139 pages, 1995.

Vol. 101: H.H. Andersen, M.Højbjerre, D. Sørensen, P.S.Eriks Linear and Graphical Models: for the Multivariate Comp Normal Distribution. x, 184 pages, 1995.

Vol. 102: A.M. Mathai, Serge B. Provost, Takesi Hayaka Bilinear Forms and Zonal Polynomials. x, 378 pages, 1995.

Vol. 103: Anestis Antoniadis and Georges Oppenheim (Editors), Wavelets and Statistics. vi, 411 pages, 1995.

Vol. 104: Gilg U.H. Seeber, Brian J. Francis, Reinhold Hatzinger, Gabriele Steckel-Berger (Editors), Statistical Modelling: 10th International Workshop, Innsbruck, July 10 14th 1995. x, 327 pages, 1995.

Vol. 105: Constantine Gatsonis, James S. Hodges, Robert E. Kass, Nozer D. Singpurwalla(Editors), Case Studies in Bayesian Statistics, Volume II. x, 354 pages, 1995.

Vol. 106: Harald Niederreiter, Peter Jau-Shyong Shiue (Editors), Monte Carlo and Quasi-Monte Carlo Methods in Scientific Computing. xiv, 372 pages, 1995.

Vol. 107: Masafumi Akahira, Kei Takeuchi, Non-Regular Statistical Estimation. vii, 183 pages, 1995.

Vol. 108: Wesley L. Schaible (Editor), Indirect Estimators i U.S. Federal Programs. iix, 195 pages, 1996.